SIX SIGMA

THE COMPLETE GUIDE TO A SET OF TECHNIQUES AND TOOLS THAT IMPROVE THE QUALITY OF PRODUCTS OR SERVICES, INCREASE PROFITS AND DECREASE COSTS, REDUCE THE PROCESS CYCLE AND ELIMINATE DEFECTS

JOSH WRIGHT

© **Copyright 2020 - All rights reserved.**

The content contained within this book may not be reproduced, duplicated or transmitted without direct written permission from the author or the publisher.

Under no circumstances will any blame or legal responsibility be held against the publisher, or author, for any damages, reparation, or monetary loss due to the information contained within this book. Either directly or indirectly.

Legal Notice:

This book is copyright protected. This book is only for personal use. You cannot amend, distribute, sell, use, quote or paraphrase any part, or the content within this book, without the consent of the author or publisher.

Disclaimer Notice:

Please note the information contained within this document is for educational and entertainment purposes only. All effort has been executed to present accurate, up to date, and reliable, complete information. No warranties of any kind are declared or implied. Readers acknowledge that the author is not engaging in the rendering of legal, financial, medical or professional advice. The content within this book has been derived from various sources. Please consult a licensed professional before attempting any techniques outlined in this book.

By reading this document, the reader agrees that under no circumstances is the author responsible for any losses, direct or indirect, which are incurred as a result of the use of information contained within this document, including, but not limited to, — errors, omissions, or inaccuracies.

Table of Contents

Introduction 1

Chapter 1: What Is Six Sigma, A General Overview And A Little Bit Of History About It 9

Chapter 2: Six Sigma Concept And Methodologies (dmaic, dmadv...) Features 20

Chapter 3: Key Concepts And Principles.............................. 35

Chapter 4: Six Sigma Belts.. 49

Chapter 5: How To Get Started With Six Sigma................... 56

Chapter 6: Tools And Areas Of Application 72

Chapter 7: Benefits 84

Chapter 8: Criticism 98

Chapter 9: Lean Six Sigma Certification 112

Chapter 10: 10 Six Sigma Do's and Don'ts 118

Chapter 11: Six Sigma Success 125

Chapter 12: How Six Sigma Can Make Lean Even More Effective 138

Conclusion 148

Introduction

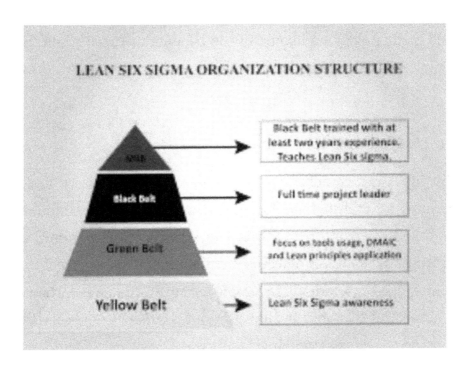

Most people believe that a company exists only to create profit. However, this is not exactly true. If you were to ask any CEO of a successful corporation why their organization exists, they would tell you that their main purpose is to create value for their constituents. In other words, a business exists to serve the owners, shareholders, and customers.

This value is created by utilizing the available resources and generating outputs that are greater in value than the inputs. This entire process chain cannot be profitable unless the business processes are effective and efficient.

So what does all this have to do with Six Sigma?

You are going to learn how Six Sigma can be used to ensure that the business process generates maximum value at a minimum cost. You will discover the philosophy and principles behind this methodology and some of the benefits incurred when Six Sigma is adopted in an organization.

Defining Six Sigma

Six Sigma can be defined as a system of statistical-based tools, techniques, and methodologies that are designed to eliminate defects and errors in products and services while minimizing process variability. A defect is anything that does not meet the customer's expectations. In simple terms, Six Sigma is a system that ensures that your business produces products and services that are of a *consistently* high quality.

If you take a good look at the business world today, you will realize that every organization worth its salt maintains an online presence. Since we live in the information age, news tends to spread like wildfire.

Therefore, companies have been forced to raise the quality of their goods and services, lest a disgruntled customer grabs their phone and tweets or posts a comment about bad service. Businesses must now guard their reputations and brands by making sure that they consistently provide products and services that meet the customer's needs.

Of all the possible quality systems available, it is Six Sigma that has been accepted as the standard.

But how does this relate to performance?

Sigma is usually represented by the Greek letter, σ, and refers to a measure of variability. If you want to determine the performance of a business, all you have to do is measure the sigma level of the processes it uses. It is important to understand that every process has an average or mean value.

Any deviation from the mean is considered undesirable, but since no product or service is perfect, some allowance needs to be made. This is why the Six Sigma process is divided into six standard deviations from the process average.

We have 1 σ, 2 σ, 3 σ, and so on until the maximum of 6 σ. Most companies usually have processes that perform at 3 or 4 σ, but the goal of every company is to attain a level of 6 σ, which is based on defects per million opportunities. The standard for Six Sigma is 3.4 defects per million opportunities. However, the good news is that even if you do not attain the 6 σ level, any improvements you make between 3 σ to 5 σ will lead to a substantial decrease in costs and an increase in customer satisfaction.

The Philosophy Behind Six Sigma

Six Sigma is a scientific methodology, and as such, you must use scientific techniques when developing your business processes and systems. It is these scientific techniques that the employees will use to improve the value of the products and services. This will ultimately benefit both the customers as well as the shareholders.

As a philosophy, Six Sigma embraces the idea that every individual business process can be measured and then improved. Here is a simplified way of looking at how the Six Sigma philosophy is used in a business:

1. The company identifies one key aspect of its business, maybe a process that is considered to be underperforming.

2. A hypothesis is developed. This hypothesis must be consistent with what is being observed.

3. The company runs some experiments to confirm their observations about the process. As more observations come to light and fresh data is recorded, the hypothesis is adjusted.

4. Statistical methods are used to separate real data from noise.

5. Steps 2, 3, and 4 are repeated until the hypothesis matches the actual results.

If the company continues to follow this system over an extended period of time, it will be able to come up with a theory or model that enables it to easily understand every internal business process as well as its customers. The end result is that instead of making decisions based on hypotheses or guesswork, management will begin to rely on hard data.

Though most organizations think that they operate on real data, the truth is that they do not. Some are simply being run based on traditions. This is why you will hear a manager say, "That's how we have always done things around here, and it works."

Yet the reality is that a company that focuses on using the Six Sigma approach systematically and consistently will improve its performance over time and leave the rest drowning in its wake.

Principles of Six Sigma

To guarantee success when using Six Sigma, you have to learn these five key principles:

1. Customer focus

Before you improve the quality of a product or service, you must first define the word "quality." The best person to define what quality means is not you but your customer. Therefore, focus on the feedback from your customers and adjust your processes accordingly.

2. Identification of causes of variations

When you are dealing with a business process, you need to understand that variation is your enemy. Your products must be consistent in terms of quality. If customers cannot trust that what they buy today will be of the same quality tomorrow, they will flee to your business rivals.

The first step in identifying the root cause of a problem is to understand how the process works. Not how it is expected to work but how it actually works. To achieve this, you need to:

- Identify the kind of data you need to collect

- Clearly define why that data must be collected

- Clarify the information that the data will reveal

- Communicate the terms clearly

- Make sure that the measurements taken are precise and repeatable

- Develop a standardized system of data collection

Data collection is generally done by interviewing various stakeholders, observing the process, and asking the right questions. Examples of questions that you should ask include:

- What can we do to make your job easier?

- Why is this process done this way?

- Are there any tasks that you perform that seem unnecessary?

Once you have collected the data, use the information to find the underlying cause of variation.

3. Elimination of variation

The next step after identifying the root cause of variation is to eliminate the variation. To achieve this, you have to make adjustments to your business process. These adjustments include eliminating any steps that don't add value to your customer. The end result will be the elimination of defects and minimization of wastage.

4. Teamwork

To ensure that Six Sigma is implemented properly, you need to have a diverse team that is committed to incorporating the Six Sigma methodologies. When a team has members with diverse skills, they will be able to detect variations in the different areas of the business process.

The team must be highly skilled and trained in the use of Six Sigma tools and techniques. There are five levels of Six Sigma certification. These are Master Black Belt, Black Belt, Green Belt, Yellow Belt, and White Belt. The highest certification is the master black belt.

5. Flexibility and thoroughness

To successfully implement Six Sigma in an organization, the system must be willing to accept change. Management and employees must be flexible in their thinking so that they see the benefits of the changes being implemented. This means that the employees must be clearly informed about how the changes will affect their work.

If they are consulted early, Six Sigma will be readily accepted. The implementation itself must not be so complicated that people would rather stick working with a flawed process than move to the new one. You also need to make sure that you thoroughly understand every area of the process.

Chapter 1: What Is Six Sigma, A General Overview And A Little Bit Of History About It

A frequent buzzword, or words, Six Sigma, is well-known throughout many industries. It's a highly prized certification that immediately garners respect from top-level management. So, what exactly is it? Depending on whom you ask, Six Sigma is a program, a metric, a methodology, a system, a concept, and a management tool. Any single answer isn't quite right, although all are in some capacity correct.

Six Sigma is a management system, which relies on a few core concepts and a proven methodology. Within the Six Sigma program, the people involved will gain an understanding of the purpose and use of many management tools and cultivate out-of-the-box approaches to problem-solving. Finally, Six Sigma refers to a metric, which is how everything related to Six Sigma started.

Sigma, from the Greek alphabet, makes frequent appearances in statistics calculations and mathematics. Why? When in the upper case form: Σ, the sigma symbol represents a summation or shorthand for a specific equation. In its lowercase form: σ the symbol represents standard deviation, which is particularly important in statistics and makes the mathematical concept applicable to daily life. The idea of mathematics plays a much more significant role in Six Sigma practices than the actual use of calculation and measuring out sums. It's particularly fun to note that 6σ represents 99.7% deviation, which is the overarching goal of Six Sigma and how the program earned its name.

Within Six Sigma, you will use particular tools in various applications such as with the summation notation of Sigma in its capitalized form. You will also rely on standard deviation as the overarching goal of Six Sigma practices. The name of the system or program, Six Sigma, refers to the standard deviation of 3.4 deviations to each million. Essentially, you're reaching for perfection. In layman terms, 3.4 deviations to each million, means that for every million products created, all except four or fewer must be unflawed.

You will be doing everything you can to ensure that each step of your process is done the same way every time - from how you place orders in a large, computerized system, to how your secretary stores their files. Six Sigma can disrupt, reconstruct, or recast your entire business processes. Does it seem unrealistic that a business owner, c-level executive, or high-level manager would encourage something that is disruptive and could result in radical changes? It does, many people not in upper management or who are generally unfamiliar with Six Sigma and the success stories are wary. Imagine someone you don't know coming into your workplace and changing seemingly unrelated things, but the fact is that it happens. It happens because something is not working in the current system, and the people involved are too close to the problem to see it.

People certified in Six Sigma approach problems with fresh eyes and a new point of view. That doesn't mean that everyone involved is on board. However, when Six Sigma practitioners come into a business, they usually have the full support of top, high, and even mid-level management. They often have the freedom to implement necessary changes and help to ensure these changes stay in place and produce the desired results. However, these people don't just waltz into a business and turn the office on its head, no. Part of the training and certification is understanding buy-in, process management, and the long-term effect of getting people to want to change.

You simply can't sum up Six Sigma in a few sentences. From a high-level view, Six Sigma works to ensure that your company is 100% successful 99.7% of the time, the result of a Six-Sigma is standard deviation equations. From a close-up view, Six Sigma sometimes creates small, but always meaningful changes to control processes, eliminate waste, instill proven management methods, and more.

Six Sigma generally promises businesses that using their certified people, tools, and implementing the suggested changes will result in:

- Reduced process variation – or greater consistency

- Improved customer satisfaction – because of greater consistency

- Reduced costs – because of reduced waste and improved satisfaction

- Increased revenue – improved customer satisfaction leads to growth and reduced cost increases margins

Other benefits of Six Sigma can include boosted employee morale and more competent or capable management teams. Six Sigma is a process that will benefit all staff over a very long course of time.

Six Sigma Success Story

To understand what Six Sigma is, and the possibilities of its impact, you must see it in motion. Throughout this book, you will see numerous success stories from Six Sigma projects. The first will hopefully hit home for many people.

The Akron-Canton Regional Foodbank regularly accepts donations and then provides the donated goods to those in need. Before 2016, before two of their staff received Six Sigma training and certification, donated goods would sit in the donation center for an average of 92 days. That is over three-months that the food was possibly going to waste, or perhaps expiring. It also meant that people in need weren't getting the help Akron-Canton was trying to provide.

After their staff received training, they began looking at the processes. Akron-Canton is a non-profit organization. They don't create products. They don't even rely on a regular shipment from a supplier. From the outside, it looked as if there were no aspects of their processes that they could control. However, they reduced their holding time from 92 days to 39 days, and the team continues to work toward the goal of reducing that to 20 days.

Within the organization, they incentivized teamwork and boosted the positive impact that volunteers felt through participation. They created clear and unwavering processes for sorting, inspecting, packing and delivering food using nothing but the Six Sigma principles, methodology, and tools.

Origins and Expectations

Where did Six Sigma start? This question is one of the most popular inquiries that newly interested people have about the program. It's often followed with what to expect during training or as a business. As you learn where Six Sigma began, it will quickly become clear what to expect, both as a trainee or as a business, looking for a Six Sigma expert.

In 1986, a board member, along with an engineer and a psychologist, within Motorola, developed a program to help eliminate waste in their supply chain. The new take on supply chain management and monitoring began Six Sigma. Motorola was on board with Bob Galvin, Mikel Harry, and Bill Smith's ideas to improve quality, reduce defects, and ultimately have happier customers. At the time, Motorola was beginning its relationship with China led by Robert Galvin, then Chairman of the Board.

They were also developing pagers and reducing their products by eliminating lesser-used lines such as car radios. For the company, this was a time of massive change, many companies could not have navigated this much change without effective control from the top down, and compliance from the bottom-up. Six Sigma has many well-founded beliefs in handling organizations during times of change.

Those who come into a company with a Six Sigma background will often incite change as a means to accomplish business goals and improve the quality of the products at hand.

From the start, Motorola acknowledged that the Six Sigma program was in development and in use. They did not set out an entire plan that would be a one-size-fits-all solution. Instead, they turned their immediate focus toward aspects that all manufacturing supply chains shared. From 1986 to 1990, the developers of Six Sigma worked and reworked the Six Sigma program to identify elements in the manufacturing aspect of business controls. Known as the Manufacturing Age for Six Sigma, the only goal was to improve quality initiatives, reduce errors, and to implement metrics for quality.

During the Manufacturing Age, Mikel Harry and Bill Smith designed the classic four-phase method system used today. The MAIC methodology is part of Six Sigma teaching at every level. Through this, Six Sigma training became the prime method for teaching business professionals to measure, analyze, improve, and control nearly every aspect of their day.

After 1990, the Financial Age began. It saw the start of Six Sigma leaders starting to look past the controls within their factories. In 1990, Motorola began to garner attention from other mega-firms or super-companies such as Toyota. Working with Texas Instruments, Motorola led its first company through learning and implementing Six Sigma methodology. This interaction would set the stage that Six Sigma would be taught and then adopted by a company, rather than an ongoing position or a consulting instance. Through the Financial Age, many companies began turning to Motorola for training in an effort to turn around their operations or to better manage their resources.

Moving into the mid-1990s and beyond are the Refinement and Adoption Age. In the mid-1990s, a critical person learned Six Sigma and became its most fervent supporter. The CEO of General Electric, Jack Welch, learned of the program and invested time as well as money to implement the tools throughout all GE divisions. GE worked with the Six Sigma leaders of the time to further develop strategies and create more succinct definitions. They build the service operations programs and tools together and expanded MAIC to DMAIC.

These tools sound foreign for now, but soon you will learn what these abbreviations mean and how they are critical components of Six Sigma.

The key takeaways from the origins of Six Sigma are that it takes a look at the business from a new angle. It took one person at the very top and one person who fully understood design and development to notice that they were missing something important. When Smith and Galvin came together to build a foundation, they weren't looking at cutting corners or evading the law. They wanted instead to see how leadership could affect production, how production could impact quality, and how quality impacted the bottom line. They saw a very straight line between warehouse manufacturing processes to the end-user. Unfortunately, even today, managers, directors, and chief executives are taught to worry about their own department. Six Sigma goes beyond that, and its progression makes that apparent. Motorola could have easily kept this methodology to themselves. Instead, they sold their Six Sigma services again and again to companies that would benefit and strive because of their teaching.

They focused on improving total quality through changing processes, culture, and leadership practices. So, what should you expect? No one should go into Six Sigma believing that it will be an easy certification that will land you a ton of jobs. You should expect to be told that your current systems, processes, and even habits are wrong. How you address people or approach change may not fit into the Six Sigma principles. Many people struggle with that as they go through the training program for any of the belt levels. After the training though, you will have a much broader perspective and with the tools you need to assess and implement necessary change. What's more, Six Sigma teachings can impact a company from any level. Yes, in some cases executive-level support is essential. However, anyone can take the basics of Six Sigma and make drastic improvements to their immediate work environment and culture.

Companies should expect changes, as well. Often businesses will bring in Six Sigma experts to implement one change and then realize that the one change won't produce the results they were expecting. Any business leader or owner should know that there are possibly a thousand ways to accomplish their goals and that Six Sigma trained staff are there to help guide the business toward higher quality, fewer defects, leaner operations, and improved culture.

Are these expectations met every time? There are very few cases where Six Sigma operations fail, and often it does resort back to lack of support. This margin for error attributes to that .03 percent that aligns with Six Sigma. Human error is always a possibility. The possibility of human error is one of the primary reasons why Mikel Harry, the psychologist and co-founder of Six Sigma was so involved. As a psychologist, he deconstructed the culture within Motorola at the time. Training and retraining leadership staff was a vital aspect of developing the methodology and making it a success. Without Mikel Harry Six Sigma would not have the impact that it does now.

The Current Age of Six Sigma is more about tailoring it to the few changing aspects of manufacturing and finance management. However, someone with a twenty-year-old certification has equal qualification to someone certified within the last year. When initiating an experience either through learning Six Sigma or hiring a certified person, keep an open mind. That is the only advice that is useful in any capacity. Be willing to accept that things must change, or things will stay the same. In that respect, Motorola developed this methodology in a time of significant change for them. It was probably alarming and unnerving to not know if any of their significant shifts in company focus would produce desirable results. They persevered and won, in that they set the stage for many businesses also undergoing or in need of change.

Here there were many success stories mentioned as you are probably well aware of Motorola, Texas Instruments, and GE. To provide a bit more tangible evidence as to the success that companies experienced and the possible futures that any of these super-companies may have experienced, this needs further explanation.

Within seven years of creating Six Sigma, Motorola saved approximately $1.4 billion in only manufacturing costs. Those cost reductions if not put into place, could have crippled Motorola. General Electric, GE, acknowledges an annual benefit of more than $2.5 billion because of its Six Sigma tools and methodologies.

Chapter 2: Six Sigma Concept And Methodologies (dmaic, dmadv...) Features

Steps to Follow in the Six Sigma Methodology

While some companies decide to start from the very beginning with a new process in the hopes of getting fewer defects and providing better service for their customers, this is not always something that needs to be done. It is likely that your business already has a process in place. Moreover, with a few simple tweaks, you will be able to make the right changes that can make it Six Sigma compliant. This is why most companies will use the DMAIC method to help them with Six Sigma.

While this may seem logical, you may be asking how you would do this in a way to ensure you did it right the first time. The steps in this methodology are going to ask you to adopt a simple five-stage process, which includes:

- **Step 1: Define**

For this first step, you are going to look at the process or the data and find the area or the process that needs improvement. This is the nature of your problem. During this step, you will also form a team and help them train in the Six Sigma method so that they can work with you to improve the current process. You must make sure that you pick out a team that is motivated and believes that Six Sigma is important, otherwise things could get messed up a bit.

The next thing to do is to identify the customers or the people who would be the most impacted by this project. You can also document the critical requirements for these customers. Then you can create a team charter that is going to detail things like the business case, the project scope, and the statement about the problem. This is going to help you finish up the define step.

There are some different tools that you can use during this phase to make it a bit easier. Project Charter, SIPOC, and stakeholder analysis are all good diagrams that can help you to depict the different elements of your new project visually.

- **Step 2: Measure**

You will find that the measuring step is going to take a bit more time compared to the defining stage. It is during this particular stage that you are going to define what parameters you will use in order to measure how to see whether performance has improved. You will also define the baseline performance as well as the extent to which the process can be improved.

You may spend some time in this phase looking at and identifying the key defects in the process. Then when you define the key measures to improve, the data is going to be collected so that you can analyze the differences between the desired performance that you want and the current performance that you already have. You should also take some time to establish the process variations during this phase.

- **Step 3: Analyze**

The third step that you are going to work on is to analyze. During this particular phase, the data that you collected in the past phase will be used in order to analyze what the gap is between the current and the desired performance. You can then do a root-cause analysis to help you determine what can be causing the gap between your current performance and the goals that you want to reach. This is often going to be calculated using financial terms so you can see the issue in dollars.

- **Step 4: Improve**

Now that you have had some time to define the problem, measure the gap that is there with performance, and then analyze the reason for this gap, it is time to move over to the fourth step. During this one, we are going to take the steps that are necessary in order to improve the issues that are there.

During this particular phase, you are going to devise a set of solutions that you could use. Sometimes there may be only a few options to work with, and other times there may be many different options from which to choose. From there, you can pick out the best possible solution based on the Six Sigma method and the options you have decided on.

The main outcome that you want to get out of this phase is to design a performance improvement plan. This plan is supposed to work in order to provide you with a measured difference in your existing process so that you can really see your defects go down.

- **Step 5: Control**

The final phase that you can work with in the Six Sigma methodology is the control phase. Here, you will come up with the project management plans as well as any procedures that need to be followed in order to sustain the new process you created.

Many companies forget this step, but it is very important. How is the Six Sigma method going to work if no one follows the plan that you come up with along the way? During this phase, you will need to document the process that you revised, devise and then deploy your response plan, and then transfer this information about the new process over to the management and others who need to use it.

When all of these parts come together, you will find that the Six Sigma method can work really well. It helps you to figure out where the problem areas of your company are and can provide you with data in order to come up with a plan to fix them. It can be time-consuming and you need to make sure that everyone in your company is on the same page when it comes to using Six Sigma so you can get the best results out of it for your company.

Now that we have some of the basics down about working with Six Sigma, it is time to help you get started with a project using Six Sigma. The best way to learn how to use this process is to actually get a project and get to work. You can read about it all that you want, but it is hard to understand until you get a project in hand and can get started on it.

- **Scope the project**

So, a Six Sigma project is going to start out as a practical problem that is affecting a business adversely. Then it is going to end as a practical solution that can help to improve how a business is able to perform. If you can find a project where a business needs some help with their current processes, then you may have a good option for Six Sigma.

The focus of your project with Six Sigma is going to be to solve a problem that could be hurting some key elements of performance for a company. Some examples of these could include:

- Process capability
- Costs
- Customer or employee satisfaction
- Organizational viability
- Revenue potential
- Cycle time
- Output capacity

You will want to get started on the project by stating out what the problems with performance are. Make sure you use terms that are quantifiable and that are going to define expectations. These also need to relate to the levels of timing and performance that you desire in the end.

As you are going through and defining the project that you want to use, you should also pay attention to some important issues. For example, some are going to be especially good for warranting a Six Sigma level of effort. Some problems that you can consider include:

- Issues that are going to give you results that seem to exceed the effort that is required in order to see improvement by quite a bit.
- Not easily or quickly solvable if you use some of the other methods that you have seen in the past.
- Issues that are going to improve the Key Performance Indicator by more than 70 percent over the existing levels of performance where they are now.

There can also be a type of flow that comes to your project as well. You will want to flow in the order below to help you work on a Six Sigma project in the correct manner:

- Practical problem: This is a chronic or otherwise systemic problem that is affecting the success of your process.

- Six Sigma Project: This is going to be an effort that is well defined and that states your problem out in terms that are quantifiable and which have known expectations.

- Statistical problem: This is a data-oriented problem that will use data and facts to help figure it out.

- Statistical solution: This is again a solution-driven by data and has known risk and confidence levels. This is in comparison to "I think" solutions that may have been used in the past:

- Control plan: This is a method that you can use that will assure the long-term sustainability of your solution for the problem. You do not want to come up with a solution that seems to work today, but then does not work a few months or so down the line. You want to go with a solution that can work for in the long-term and still provide you with some great results.

- Practical solution: This type of solution is not seen as irrational, expensive, or complex. The best part is that it can be implemented without a lot of problems or a wait time.

- Results: These are the tangible results that you get. You will be able to measure it out financially or in other ways that show how it is benefiting the company.

Getting started on a new project can sometimes be the hardest part. You want to make sure that you are working on a project that can help the business out, and that will let you use the Six Sigma methodology to get it done.

Six Sigma is one of the most effective methods currently available to help improve the performance of any organization. It is able to do this by minimizing the defects in a business's products or services. With this method, all the errors committed have a cost associated in the form of losing customers, replacing a part, waste of material or time, redoing a task, or missing efficiency. In the end, this could end up costing the business. Six Sigma works to reduce these losses in order to help a business grow.

This methodology of Six Sigma was endorsed by Motorola in the 1980s. The company, at that time, was trying to find a way that they could measure their defects at a granular level compared to the previous methods, and their hope was to reduce these defects in order to provide a better product to their customers.

What they ended up with, was a huge increase in the quality levels of several of their products, and the company received the first Malcolm Baldrige National Quality Award. It did not take long until Motorola shared their Six Sigma method, and soon there were many other companies who were reaping the rewards as well. By 2003, it was estimated that the combined savings of all companies using the Six Sigma method were more than $100 billion.

DPMO

The term defect is used a lot when it comes to Six Sigma. The goal of the company is to reduce how many defects occur so they can reduce waste, provide a better product for their company, and make more money. However, what does this defect mean?

The "defect" is going to be explained as the nonconformities that are showing up in an output that falls lower than what the customers find as satisfactory. The number of DPMO, or defects present per million opportunities, is going to be used to figure out which part of the Sigma scale that process corresponds to. Most organizations in the world would fall somewhere between Three and Four Sigma. This may not seem so bad, but it really implies that they could be losing up to a quarter of their total revenue simply because there are some defects in the organization.

The Six Sigma methodology can help these businesses out. It can move them up to a new level of Sigma, which can reduce all that waste and those defects, effectively helping them to earn more profits.

Applying Six Sigma

While there are many different methodologies that can come with Six Sigma and can help the business to reduce its defects, the two basic ones that are good to start with include DFSS and DMAIC. Let us look at each of them below to understand how they work a little better.

DMAIC

The first one is DMAIC. In order to help modify a process that is already in existence and change it so that it can be more compliant with the Six Sigma methodology, therefore, more efficient, you would work with DMAIC. This stands for:

- Define: This is where an organization needs to define the goals for process improvements so that they are in coherence with the strategies of the business and with the demands of the customers. You can't get far in your process without defining what goals you want to reach and which processes must be improved to reach these goals.

- Measure: This is the current performance of the systems in place for the business. It will also take some time to gather data that is relevant and can be used in the future. Measuring the data you receive and the results that you are looking for can be important, and you must make sure you are relying on the right tools to do it.

- Analyze: This is where you will analyze the current setting and then observe how the relationship between the performance and the key parameters work. Lean analytics can be a good tool to use to analyze the situation and make sure your improvements are actually working. If changes need to be made, your analysis will showcase when this should happen.

- Improve: From the other steps, you will be able to find ways to improve the process. This helps to optimize the process to earn the business more money.

- Control: Here you will control the parameters before they are able to affect the outcome.

Differences between DFSS and DMAIC

1. DMAIC is used when improving a process that is already in place while DFSS is used when developing an entirely new process.

2. DFSS is considered a preventative approach rather than a curative one. Organizations implement DMAIC methodology only when they identify flaws in the process and try to eliminate waste. With DFSS, the defects are eliminated while the process is being designed.

3. DFSS is considered to be more economically viable than DMAIC since defects are eliminated during the initial stages of process/product/service design.

4. The tools used in the implementation of DFSS are quite different from those in the DMAIC methodology. In fact, the reason why DFSS was created was because DMAIC tools could not be used to optimize a process beyond three or four Sigma without having to redesign the fundamentals. The best option was to design for quality from the start.

DMADOV

This is a slight modification of the DMADV methodology. It contains an *Optimize* phase where advanced statistical models and tools are used to optimize performance.

DCCDI

Define – Definition of project goals.

Customer - Completion of customer analysis.

Concept – Development, review, and selection of ideas.

Design - Detailing how customer needs and business specifications are to be met.

Implementation – Development and commercialization of the product or service.

IDOV

This methodology is well known in manufacturing circles.

Identify – Finding out the customer CTQs and specifications.

Design – Customer CTQs are translated into functional needs and further into potential solutions. The best solution is chosen from the list.

Optimize – Advanced statistical models and tools are used to optimize performance.

Validate – Ensuring that the design will satisfy customer CTQs.

DMEDI

Define – Identifying business problems and customer desires.

Measure – Customer needs and requirements are determined.

Explore – Analysis of the business process and exploring options for designs that will meet customer needs and specifications.

Develop – Delivering an ideal design as per customer needs.

Implement – Putting the new design through simulation tests to check efficacy to meet customer requirements.

It is evident that DFSS encompasses numerous methodologies. There are variations in the names as well as the number of phases. However, all the methods under DFSS utilize the same advanced techniques for design, for example, Failure Modes and Effects Analysis, Design of Experiments, Robust Design, Quality Function Deployment, etc.

Chapter 3: Key Concepts And Principles

- **Key information**

- **Names:** Six Sigma, 6 Sigma, 6 σ

- **Uses:** a qualitative, quantitative and structured approach to business management.

- **Why is it successful?** It is a precise approach to improve key business processes for a reliability of more than 99.99%. The objective is to achieve an average of 3.4 defects per one million defect opportunities (where 3.8 sigma, for example, corresponds to 10 000 defects per million).

WHY IS CASH A TRUE COST TO ANY BUSINESS??

Before getting into the detail of the 7 Wastes, I feel it's important to talk about CASH and the cost of CASH within a business – I mention it a lot throughout this guide.

Please understand that I am not an Accountant, so I will explain the principles of the costs of cash in the only way I know how - as an Operations Manager.

My explanation may not be as technical as an Accountant's but I am sure it will be understandable! If you feel you need a more in-depth definition, I'd recommend you ask the Accounts team in your business – or failing that there are some great websites out there!

Looking back now, I realize that in my early years working in manufacturing I really didn't understand why we just didn't buy more raw materials to make sure we didn't run out or why we didn't plan in much longer runs to get really good factory efficiencies.

At the time, I didn't understand that lack of profit is a serious problem for a business but that lack of CASH is very rapidly fatal. A business can be very profitable but still fail if it runs out of CASH.

CASH is the tightrope walked between the paying out of money for raw materials, wages, rent and all of the other outgoings and the paying in of money from customers.

If, for example, you have a raw material supplier that demands to be paid after 30 days but your customer (let's say it is a large and powerful supermarket!) only pays you after 90 days you have somehow got to find 60 days worth of money to keep your business afloat!

Generally, unless the business has a generous owner or is well established and has lots of cash reserves, it will be borrowing money or will be factoring invoices.

It's well known that when you borrow money you have to pay interest to whoever has loaned it to you. So I won't need to explain how the interest you pay on the money that you have tied up in your business in raw materials or stocks is a true cost to the business.

Factoring is slightly different but is a real cost too. To help cash flow some businesses effectively "sell" their invoices to a finance company.

How it works is this:

- I sell £10,000 of bread to a supermarket. I need this £10,000 to pay my staff, pay for the raw materials I used to make the bread and keep the lights and heating on. I can't wait for 90 days to be paid!

- I approach a finance company that gives me £9000 now and they collect the £10,000 from the supermarket in 90 days time.

- Whilst this has solved my cash-flow problem it has cost me 10% of my sales revenue.

As you can see above, with factoring or with borrowing, there is an actual cost when you have money tied up in stocks and inventory.

There is another way to consider this too – if you have hundreds of thousands of pounds lying on shelves in the form of stocks, this is money that is not available to you to improve the business and make it more efficient and profitable in the longer term.

I would urge you to get your head around this principle if you haven't already "got it". It is fundamental to your understanding of the 7 Wastes and it will hugely benefit your career if you are able to think like an accountant or a business owner.

HOW IT WORKS

Equipped with this different way of looking for activities that don't add value, you'll be able to identify <u>where</u> and <u>when</u> in your process there are one of the 7 Wastes (actually there are 8…more about this a bit later!).

Once you've stepped through each part of your process, production line or even full factory you will have a comprehensive (and long most likely!) list of newly identified wastes.

It's easy at this stage, especially if this is your first attempt at it, to be intimidated by the very large amount of items you've just written down in your list.

By the way – if your list is very short then I'd suggest that you either have one of the world's most process efficient factories or you've probably not looked hard enough! I don't say this as a criticism – all factories and plants will have in-built processes that add little value and need to be classed as waste. Examining the 7 Wastes is a tool to shine a light on all actions which are less than 100% efficient – it will by necessity throw up a <u>huge</u> list. Some of which though you won't be able to remove or the costs of removing will be too high.

So – don't worry about how long the list is (unless it's not long enough).

The next step is to try to put a cost to each one of the wastes you've identified – a rough cost may be all you can calculate if the waste is quite hard to measure like Non-Use Of Skills. With some wastes like Defects an accurate cost measure would be really quite easy.

Once you have a cost – if it's accurate or a rough estimate – you can start to rank the big wins and draw up an action plan to reduce, remove or eliminate as many of these wastes as you can.

As usual - you will need to make sure that the amount of investment needed is outweighed by the amount of savings you are going to make! Some of the wastes you identify will save a little and cost a lot – sensibly, (unless there are social, safety, moral or environmental reasons) these wastes will not be worth engineering out!

Remember – you can't win them all....

GETTING INTO DETAIL..

The Lean Manufacturing term "7 Wastes" is well known but I would argue that you should actually consider "8 Wastes" when you are thinking about your plant and your process!

I'll continue to refer to these as "the 7 Wastes" but will actually explore 8 different wastes through this guide.

On a side note - there being 8 wastes instead of 7 allows you to use a memorable acronym to help you easily remember the wastes: **DOWNTIME**.

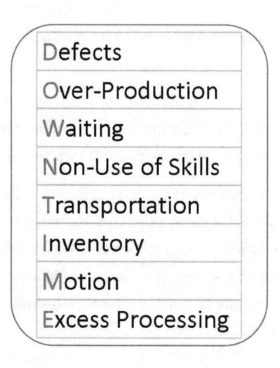

THE "WASTES"

Defects

Definitely the most visible of all of the Wastes.

At it's most basic this covers all material/products/components which are unusable as they are due to mistakes or errors in the process. These would have to be binned or reworked.

But the effects of Defects goes much further than simply counting the number of unusable components that have ended up in the skip or down the drain.

For every Defect there are further consequences and costs.

- Faulty goods that reach a Customer results in complaints, refunds and potentially a Customer lost for good.
- Defective goods that are discovered before reaching a customer need to be reworked or replaced.

Over Production

More often than not a Production team member would consider Over Production to be impossible! Beating a production target or making a large batch size which gives the team a long efficient run is what Production teams dream of!

But - Over Production is one of the most common and costly of the Wastes as it is frequently overlooked as being an actual "waste" and it hides a large number of underlying problems within the process.

The need for a long run to drive efficiency will be a sign of a number of issues such as lengthy changeover times, excessive equipment downtime, poor setup, process inconsistency and lack of process flexibility. These need to be recognized for what they are and addressed.

It's not only the underlying problems in the factory that can drive Over Production into your process. Unreliable suppliers, erratic sales patterns and an inflexible supply chain can all be responsible for making a business take an approach of "JUST IN CASE" rather than "JUST IN TIME".

We need to agree that although Over Production can be helpful to smooth out efficiency loss through changeovers and provide a buffer between you and unpredictable customer sales, it does drive excess stocks of goods and materials not immediately needed into the Warehouse.

This absorbs storage capacity, plant flexibility, responsiveness to customers' needs and also cash.

Waiting

Reasonably self-explanatory as a Waste - the amount of time lost because two or more dependent processes are not balanced and synchronized. This could be expressed through waiting for a process to start, to finish, or waiting during a process.

There can be a number of root-causes of this - it could be due to poor communication (e.g. waiting for a decision from somebody!), inefficient and lengthy changeovers, unbalanced processes where one component is faster than another, or where a process upstream is delayed.

A classic example of this within a manufacturing business is where a supplier (internal or external) fails to deliver at the required time. This causes all downstream process to stop and wait!

Non-Use of Skills

Not always included in the Wastes (this is the 8th Waste!) because, while it is a real waste within a business, it's hard to quantify the cost.

Every team and every department will have colleagues and team members who have untapped potential or skills that are unknown to you (sometimes unknown to themselves). They will also have a detailed working knowledge of the process as they are the ones working "at the coal face" for 8 hours a day.

Whenever attempting to solve an existing problem, to reduce defects for example, discussing the project with the wider Operations team will reduce the potential to miss ideas and skills sitting unused within the team.

Insight and involvement from within the team, if looked for, can make a considerable difference when it comes to driving out the other Wastes. When looked at this way, you can see why not using this potential source of skill, insight and help is a Waste.

Transportation

Transportation is the movement of materials and people around your business and, if not part of the actual value-adding process, it is adding cost and creating waste.

Using Fork Lift Trucks, conveyors and people add cost to your process - equipment has to be purchased and maintained and people have to be paid. There is also the hidden cost of loss of time - while the product or material is being transported it is not available to have additional value added, so this is a lost opportunity.

A major cause of transportation waste will be poor factory or site layout. Poor flow where sequential processes are not physically close to each other results in unnecessary movement between steps.

Multiple storage locations due to insufficient space can be a factor to consider, as can over complex processing operations which may involve lots of movements.

Noting down each step of the process and how much transportation is taking place (this could be done in distance or minutes) is the first step to begin reducing the Waste of Transportation.

There may be a limit as to how much of this you can drive out without a significant cost as many older plants who's use has been adapted to changing circumstances over the years can find that they are transporting materials between processes through a veritable rabbit warren.

There will be things you can do to improve the situation but solving all Transportation waste may involve demolishing the plant and starting again! Hardly a realistic solution...

Inventory

Stock inventory that is not in use or sold to the Customer is Waste.

There will be significant cash tied up in raw materials in Stores waiting to be processed or in Work In Progress (WIP) within the plant.

Using Just In Time principles to reduce the amount of stock inventory will significantly improve your business's cash flow.

Many people do not regard this as being an actual waste - however there is a real cost involved in having racking packed with materials in Stores.

All inventory has to be paid for by a business and there is a genuine "cost" of cash to every business. The business may be borrowing money from a bank at a cost of interest, could be factoring invoices at a rate of 10% or could perhaps have used that it's own cash to improve the efficiency of the factory rather than have it tied up in materials.

Motion

Excess Motion is a commonplace Waste - some of which can be designed or engineered out, much of which is unavoidable and is part and parcel of any manufacturing business.

It is the movement of people or equipment more than is absolutely necessary to accomplish the task in hand and it is different from excess transportation in that this waste needs to be considered in the context of the negative effect it has on your plant or people. People tire quicker and machines wear out faster with excessive and unnecessary motion.

At it's most simple - it may be a person who is required to twist their body to pick up something that could have been presented in front of them, someone needing to bend down to a low conveyor belt that could have been presented at waist height or it could be machine part moving through a longer stroke than needed to complete the operation.

Reconfiguring storage locations, work stations, conveyors with the waste of motion in mind can eliminate much of this (using the principles of ergonomics will support this process).

It is unavoidable to have machine parts wear out and should be considered as inevitable – but reducing excess motion will help to extend life and reduce costs associated with breakdowns, servicing and (in the long term) replacement.

Excess Processing

The Waste of Excess Processing takes me right back to my early days in Manufacturing when I was asked by Factory Manager to explain to him which of two pens was the "Quality" pen.

He presented me with a clearly very expensive fountain pen and a cheap, throwaway ballpoint pen and asked me to choose which was the Quality item.

Being young, new to manufacturing and relatively "green" I answered that the finely crafted fountain pen was the good Quality one.

As you can guess - this was a trick question by my Factory Manager designed to teach me an important lesson. The answer is that they are BOTH quality items as they are both designed with a specific customer, pricepoint and use in mind.

Excess processing is the concept that adding more quality to a product than the customer actually wants or is prepared to pay for is a waste.

In the context of my illustration above - a Supervisor such as myself at that time would not have been prepared to pay for the expensive fountain pen for use in my day-to-day work in the plant, while the Factory Manager may have felt that pulling out a throwaway ballpoint pen to sign a significant contract with a customer did not convey the right image.

In the automotive industry - painting a car body part that is going to be hidden within the engine compartment and never seen is Excess Processing, over buffing and polishing and cleaning to a higher standard than is required are also examples of adding more cost than is needed.

A frequently un-regarded form of Excess Processing is created by having specifications tighter than are required by a customer as this will drive waste through products rejected as quality defects.

If for example you are trying to achieve a specific size tolerance of +/- 1% variation, you will expend cost tightening your process to achieve this and will create defect wastage of any product outside this tolerance.

If the customer would be satisfied with a tolerance of +/- 2% as this kind of size variation would not cause any problems at all (which of us would notice or complain if a toothbrush, a banana or a bag were an extra 1% larger or smaller than expected?) then driving towards this tighter tolerance is unnecessary and a waste.

Chapter 4: Six Sigma Belts

The Six Sigma belts are classified into four, namely, the Yellow Belt, the Green Belt, the Black Belt, the Master Black Belt and the Champion.

You should know that each of these roles is meant to have tutoring, and in a lot of cases, the right certificate for the roles.

In the beginning part of Six Sigma, each of the organizations that were built had its way of doing things with regards to methodology and the techniques and tools. However, the majority of these organizations currently hope on an independent certifying body for tutoring and presentation of an approved certificate. The American Society for Quality and the International Association of Six Sigma Certification are the two most popular and recognized organizations that offer certification.

Also, the GoSkills Six Sigma courses are in tune with the IASSC

Body of Knowledge. Outlined below are the four Belts associated with Six Sigma.

Champion

Champions are the guarantors of the project. They help the Black Belts to select improvement projects to work on, estimate their potential and assess the company's products compared to those of the competition. The role of Champions is to ensure the supervision, support, and financing of Six Sigma projects and manage the staff needed to implement them. They are the pillars of the project and that is why they are chosen from the best people.

Six Sigma Yellow Belt

An organization can possess a lot of Yellow Belts. The people with the Yellow Belts are known to be team members on a Six Sigma project guided by a Green Belt or Black Belt.

They are meant to get to know the planned methodology as well as the use of cross-functional techniques and tools.

They will join in all meetings and gatherings of the project team while performing their duty of the subject matter expert for their task. This duty is carried out together with their usual full-time occupation.

A project will have a lot or some Yellow Belt members, which will be required based on the extent of the process being examined and the type of the problem.

The teachings for a Yellow Belt usually focus on the structure of the methodology as and make use of a cross-functional problem-solving technique and tool.

It is normal for an individual who has Yellow Belt certification to be part of various Six Sigma project teams.

Six Sigma Green Belt

An organization will possess various Green Belts. The Green Belt duty is usually that of a project leader. The Green Belt is normally effective on Six Sigma projects that would be around their area of duty or abilities.

These people are aware of the Six Sigma methodology and organization. The individuals are also capable of applying the Lean analysis tools as well as the statistical techniques popularly used with Six Sigma.

The Green Belt individuals guide little projects or projects that focus on only one function. This duty is usually done together with another work occupied full-time.

The majority of Green Belts are guiding a project that is related to improving a few aspects of their business processes. In few cases, a Green Belt might be charged with the function of a large cross-functional project being guided by a Black Belt.

Large cross-functional projects, in most cases, have several analyses happening at the same time, and a Green Belt will guide each of their skills.

While as a project leader, the Green Belt is charged with the duty to make sure that the right Six Sigma techniques and tools are adopted at every phase of the project.

The Green Belt individuals will usually lead the presentation and dialogue of the project at the phase gate reviews. That is because the Green Belt individuals are the only one on the project which has been tutored in the Lean analysis techniques and the statistical Six Sigma tools, they will carry out analyses.

Note that the Green Belt is not known to be the subject matter expert on every part of the product or product, but they are, in most cases, the expert on few parts of the product or process. As a result, they are mandated to fetch their subject matter expertise to endure in a similar way in which a Yellow Belt works. However, the Green Belt is not likely to be an expert on every part of the sophisticated Six Sigma techniques and tools. Anytime they encounter any problems, they will look into their Black Belt for recommendation and lecturing.

Six Sigma Black Belt

An organization will, in most situations, have various Black Belts.

The Black Belt duty is that of subject matter specialists on Six Sigma for a task or site within the organization. These individuals guide large cross-functional projects and serve as guides for the Green Belts in that department or position. This is usually known to be a full-time job.

Black Belts do not only know how to apply the tools and methodology, but they are also known to be teachers and tutors for the Green Belts and Yellow Belts within the organization. A usual day will contain:

Organizing a team meeting for one project they are guiding.

Meeting with several Green Belts to check up on their progress and offer tutoring services for their next actions.

Performing statistical analysis or value stream with data from any of the projects they are guiding.

Offer instruction and lessons on the use of Six Sigma within their organization for Green Belt and Yellow Belt applicants.

Assembling with the stakeholders of the organization to converse about the projects and identify the shortcomings for future projects.

From the above, the individual is normally supposed to oversee and lead various projects at the same tile while also performing the role of a coach for some Green Belt candidates who are leading their projects. While as a project leader, it is mandated for them to arrange and structure the work. The most difficult part of those projects is to work alongside the stakeholders from multiple functions. In different organizations, the Black Belt duty is given annually or for two years so that several individuals can be an expert in all areas of the Six Sigma methodology.

Six Sigma Master Black Belt

The majority of organizations usually have just one Master Black Belt, an individual who is usually a senior person is accountable for controlling the Six Sigma scheme within the organization. Also, the Master Black Belt is a full-time position. A lot of times, this Master Black Belt gives account to the C-level champion for the Six Sigma scheme.

From a tutoring and certification angle, the person has a similar credential as a Black Belt.

However, duty and tasks are diverse.

The Master Black Belt is not controlling projects; rather, they are controlling the scheme.

The Master Black Belt is usually performing its role with senior leadership to know the number of Black Belts and Green Belts is required and which functional departments or setting should get them immediately.

The Master Black Belt usually retains a status statement on the portfolio of the Six Sigma project, the energetic ones, the full ones, and the planned ones.

As much as they are capable of entering the impact of the whole program on the organization, and they can set the improvement efforts first based upon the approach of the organization.

These people also work alongside HR to retain the training documentation of every Yellow Belt, Green Belts, as well as Black Belts in the organization.

In cases where an organization is not as large as it should be, or if the Six Sigma scheme is not large within the organization, the duty of the Master Black Belt will be taken up by one of the Black Belts in the organization.

Chapter 5: How To Get Started With Six Sigma

Implementing Six Sigma

No two organizations are exactly alike. They may exist within the same market or industry, but they face different sets of challenges. Therefore, the strategies that they use when implementing Six Sigma will vary significantly, especially if you consider their organizational culture and strategic goals.

You will learn how Six Sigma can be implemented within an organization and how to get top management to commit and support it. You will also learn how to reduce lead time to improve customer satisfaction and ensure business success.

There are generally two approaches that an organization can use when implementing Six Sigma:

- Execute a Six Sigma initiative or program
- Establish a Six Sigma infrastructure

Executing a Six Sigma Initiative or Program

This option involves training specific employees on how to apply Six Sigma tools in the workplace. It is an unstructured way of implementing Six Sigma in an organization. A few practitioners are chosen and trained on how to use statistical tools whenever they feel that it is necessary to apply the tool. If the Six Sigma practitioners get stuck, they can ask a statistician for help.

Though this approach may yield some successes within the company, these are few and far between.

The simple reason is that there is not enough consistency, and therefore, each success fails to provide the support for the next one. It is as if there is a lack of total commitment to fully implement the Six Sigma methodology throughout the organization.

An organization that focuses on implementing Six Sigma as a mere program will only change a few of their tools and introduce a couple of training classes for the affected employees. If it were to go further, it might apply these tools to a number of special projects.

However, these projects are rarely a core part of the organization. These projects tend to be low-level initiatives that have not even been endorsed by top management. If the solution that the project is meant to provide directly affects upper management, then it is possible that the project will experience a lot of internal resistance.

It is clear to see that implementing Six Sigma through tools alone is not likely to boost the bottom line or add value to the long-term, strategic goals of the organization. This approach usually leads to Six Sigma being viewed as a fashionable methodology that is only useful during certain seasons. There will be a minimal return on the investment made to train employees.

No matter what kind of extraordinary achievements are made through the use of Six Sigma tools, the benefits will not be visible to top management.

Their resistance to Six Sigma ultimately kills any attempts to bring change, and with no assigned change champion, even getting funds to finance the initiative becomes tough. Ultimately, success can only be achieved by convincing upper management to support Six Sigma implementation throughout the organization.

Establishing a Six Sigma Infrastructure

The best way to implement Six Sigma in an organization is to focus on establishing a solid infrastructure that will guide all Six Sigma projects. This goes way beyond just introducing a couple of statistical tools that are used haphazardly.

This particular option focuses on gaining top management buy-in before any investment in Six Sigma is made.

Employees are trained to use the right tools at the right time when working on a predefined project.

Six Sigma practitioners are selected and trained for a period of four months, and in between training sessions, they are given projects to help them apply what they have learned.

An organization that invests time and money to deploy Six Sigma as part of its broader business strategy will benefit more than the one that simply deploys Six Sigma tools. Here are some of the benefits of deploying Six Sigma infrastructure with the support of upper management:

- The projects deployed directly affect the bottom line, thus achieving a bigger impact

- Six Sigma tools are used more effectively, efficiently, and productively

- It provides a project management strategy that practitioners can study and improve upon

- It makes it easier for the practitioners and upper management to communicate

- Critical business processes can be understood in detail

- It helps managers and employees understand the real value of statistical tools to the organization

One of the key steps in deploying a Six Sigma project is the project selection process. It is important to select projects that will help the organization meet its strategic business goals. Six Sigma can be a useful and effective roadmap for achieving these goals.

Regardless of how an organization chooses to implement Six Sigma, the important thing to note is that Six Sigma should be a long-term commitment. This will ensure that there is an objective analysis of every element in the business process.

It will be easier to learn from past mistakes and improve on subsequent implementation plans. This will create a closed feedback loop that ultimately reaps dividends for the organization.

Overcoming Upper Management Reluctance

There are certain measures that you can take to overcome top management resistance to Six Sigma. Here are five steps to take to gain management support:

1. Choose your projects wisely – If you develop the right project selection criteria, you stand a better chance of picking projects that will have the greatest impact on the company's bottom line. This will prove to top management that Six Sigma is of value.

2. Get quick results – Make sure that the project bears fast results and generates significant returns. You will have about five weeks to show management that you can reduce costs and improve productivity by at least 30 percent.

3. Track your interim progress – Develop a work plan that specifies key milestones, responsibilities, and deliverables.

4. Gather an expert team – You may have to seek Six Sigma experts from within or outside the organization as part of your deployment team. Having qualified professionals increases the chances of project success. This will convince top management to consider implementing Six Sigma throughout the entire organization.

5. Maintain a clearly defined project scope – Make sure that the project scope is wide enough to generate significant returns but also narrow enough to enable quick completion.

Once top management sees the practical benefits of Six Sigma, they may consider a limited initial commitment or a broad-based rollout of Six Sigma throughout the organization.

Lead Time

Lead time is defined as the time from which the customer places an order to the moment the product or service is delivered. Every type of business has some form of lead time, whether it is in manufacturing, project management, software development, supply chain management, etc. The only difference is how the lead time is interpreted in that particular industry.

But why is lead time so important in business?

Because time is money!

Let's say that a manufacturer usually buys steel from a supplier. Imagine a situation where the supplier has a lead time of one month. This means that the supplier takes one month to deliver an order for steel to the manufacturer. Therefore, the manufacturer needs to maintain an inventory of one month's worth of steel to keep producing products for their customers.

Maintaining inventory means paying storage costs for the steel, so the longer the supplier's lead time, the greater the storage charges. It would be in the manufacturer's best interest to find a supplier who can supply the steel within a shorter period of time.

Components of lead time

Lead time consists of several different elements that originate from the various departments within an organization. These include:

- Preprocessing time – This is the time it takes to receive an order from the customer, understand the order, and then create a purchase order.

- Waiting time – The time an item has to spend awaiting production

- Processing time – The time it takes to produce an item

- Inspection time – The time it takes to check a product for conformity to standards

- Storage time – The amount of time an item stays in the factory or warehouse

- Transportation time – The amount of time the product is in transit from the factory to the customer

When you add up all these elements, you end up with the total lead time. One assumption that is made is that there is no inventory in storage. In other words, we assume that the product must be made from scratch.

So the challenge you have is to find a way to reduce the lead time.

How to reduce lead time

1. Eliminate non-value activities – Use value stream mapping to find those activities that do not add value to the process and eliminate them.

2. Simplify the production process – When you make the process less complex, production flows faster.

3. Improve layout – Arrange the machinery and work process such that raw materials and finished goods do not have to move long distances.

4. Document your operating procedure – Create a document for standard operating procedures so that every employee is familiar with what is required. This will enable quick learning, less confusion, and enhanced consistency.

5. Planned machinery maintenance – It is better to schedule the regular maintenance of machinery than to wait for a total shutdown that will cripple production.

6. Find backup suppliers – Create an arrangement with a group of suppliers so that in case one supplier lets you down, you can rely on another. Take time to educate your suppliers on how your operations are run so that they understand how important they are in your business process.

The truth is that customers will always gravitate toward companies that have shorter lead times. Any organization that works to reduce its lead time will ultimately improve its chances of success.

Next, you will learn the steps necessary to apply the DMAIC framework.

The DMAIC Process

By now it should be clear that a Six Sigma project is designed to measure and improve the current performance level of a business process by using statistical tools. You are going to learn more about the phases of DMAIC and the steps that are involved in each phase of the process.

DMAIC is an acronym that stands for Define, Measure, Analyze, improve, and Control. It is simply a management system that allows an organization to generate a continuous flow of improvements to projects. DMAIC helps to determine which projects should be implemented as solutions to eliminate defects and produce sustainable results.

Phases of DMAIC

Define

This is the first phase of the DMAIC approach. In the Define phase, the goal is to clarify every single element that will be involved in the project. Here are the steps involved in the Define phase:

1. Creating a project charter and scope – This is a document that establishes the boundaries of the project. Since you cannot solve all the problems in a business process, you must define the limits of the project activities before moving forward.

2. Developing a high-level process map – This map is usually created using the SIPOC (Suppliers, Inputs, Process, Outputs, Customers) format. Once the map is complete, the project team can choose one specific area and create a more detailed plan.

3. Defining the customers, their requirements, and their expectations (CTQs) – It is important to understand who your customer is and the needs that they have. You have to reach out to your customers and get relevant feedback on how to best solve the problem.

4. Developing the problem statement and goals – A problem statement is a clear description of the challenges that the project seeks to address. You need to collect data that proves the existence of a problem and whether it is of a high or low priority. The problem statement should also include customer requirements, goals, and benefits that will accrue at the end of the project.

5. Defining the resources to be used – Project resources that need to be identified include the project sponsor, change champion, the process team, and any other material/financial resources that will be required.

6. Developing the project plan and its milestones – This should include a short statement of how the project activities will be completed, by whom, and the relevant timelines. You also need to specify the communication channels to be used.

Measure

In the Measure phase, the project team reviews and measures the state of the existing process. They create a baseline that will be used to determine the level of improvement after the project. Here are the steps involved in the Measure phase:

1. Create a detailed process map – A process map that is detailed will help the project team easily identify the areas in the process that have bottlenecks, defects, or useless steps.

2. Creating a data collection plan – Before you start collecting data, you first need to define the methods you will use and the objectives of the process. What metrics will be measured? What are the measuring tools that will be used? How often will measurements be taken? How will the data be recorded?

3. Gathering data – The data to be collected must be useful in defining the problem. It must also help in

identifying the root causes and location of the problem.

4. Verifying the measurement system – The system you will use to measure the current state of the process needs to be reliable. It, therefore, must be analyzed. If you fail to validate your measuring system, you will end up collecting misleading information.

Analyze

The goal of this phase is to spot all the factors that may be responsible for the defects, waste, and bottlenecks. This will help you identify the root cause of the problem. Here are the steps involved in the Analyze phase:

1. Identifying value-added and non-value added process steps – The project team has to look closely at every step in the business process. The goal here is to identify which steps add value and which ones do not. Non-value added steps must then be targeted for elimination.

2. Identifying sources of variation – Sources of variation are simply the causal factors that may be responsible for a defect, bottleneck, or waste. This can be achieved through brainstorming and other tools. Once you have compiled a list of causal factors,

you should rank them and then investigate the major ones.

3. Determining the root cause – One of the challenges of problem-solving is that there are times when the perceived problem keeps recurring. If this happens in Six Sigma, it is because the project team solved a symptom and not the root cause of the problem. You have to identify the root cause out of all the potential causal factors. Then you have to develop effective countermeasures using the tools available.

At the end of the Analyze phase, any new information found should be used to update the project charter.

Improve

This is the phase in which a solution is identified and developed. This is only possible after all the relevant data has been collected and analyzed. The goal is to come up with a list of possible ideas and then select the most suitable one. Here are the steps involved in this phase:

1. Performing experiments - The project team investigates how the possible root causes can be resolved to improve the process.

2. Creating innovative solutions – The project team identifies potential solutions that will improve

quality, boost safety, lower costs, and enhance efficiency. Of course, the best solutions should be able to achieve all this without consuming too many resources.

3. Assessing failure modes of potential solutions – Each potential solution should be reviewed in terms of risk and possible impact. This will help the team to identify any problems that may occur when a solution is implemented. The best solution should be low-risk and have no negative impact on the process.

4. Selecting the preferred solution – The best solution is picked and then implemented.

5. Re-evaluating potential solution – The solution that has been selected is validated to confirm whether it has improved the process and resolved all issues. This is done using statistical methods, data collection, or pilot builds.

Control

The Control phase is all about maintaining the solution. The strategy is to constantly monitor the performance of the new process and how employees are responding to it.

This should never be a one-off event. All the data collected by the process management team should be compiled and used to create a manual that will be used to improve subsequent projects and teach new staff members.

Here are the steps involved in this phase:

1. Updating the process standards – Once improvements have been made to the process, it is important to document all the changes and create new standards. These updates can be in the form of control plans, work instructions, etc.

2. Implementing statistical process control – This is a way to track how the primary steps in the process are performing. The process owner or associates should be able to note any shifts in the process.

3. Verifying benefits, cost savings, and growth of profits – A process monitoring plan should be created to track and record the *long-term* performance of the new process. This plan must clarify how the new process will be measured, the frequency of measurement, the person responsible for fixing any issues, a method of documentation, etc.

4. Celebrating – Once the project is over, it is important that the whole team comes together to celebrate. Upper management should also recognize the efforts made and benefits realized.

Chapter 6: Tools And Areas Of Application

There are many tools that you can use in order to make Six Sigma work for you. These tools are there to ensure that you are providing good quality management to your business and some of the tools are so successful that they can be used outside of a Six Sigma application as well. Some of the main methods that can be used include:

5 Why's

The 5 Why's is a technique that is there to explore the cause and effect relationship of a problem. The goal of this technique is to find out the root cause of a problem by repeating the question "Why?" Each answer is going to form the basis of the following question. The 5 in the name derives from the idea that it takes about five iterations in order to resolve the problem, but depending on your particular issue, you may need to use more.

Not all problems though, are going to have one root cause. If you would like to figure out more than one root cause, this method is going to be repeated by asking a different sequence of this question each time that you use it.

In addition, the method is not going to provide any hard rules about what lines you should explore with the questions, or how long you need to continue your search to make sure you find the root cause. Thus, even if you follow this method closely, it may not give you the outcome that you want.

An example of the 5 Why's includes the following:

My vehicle is not starting:

1. Why? The battery is not working.
2. Why? Because the alternator is not functioning
3. Why? The belt on the alternator has broken off.
4. Why? The belt should have been replaced a long time ago, but was not.
5. Why? The vehicle owner did not follow the required maintenance schedule for the vehicle.

This helps to show why there was an issue with the vehicle, and you can easily choose to take it further into some more why's until you find the solution that you are looking for.

Axiomatic design

The axiomatic design is a systems design methodology that is going to analyze the transformation of the needs of the customer into design parameters, functional requirements, and process variables. The method is going to get its name because it is going to use the design principles that govern the analysis and decision-making process. The two types of axioms that are used with this process include:

Axiom 1: This is the independence axiom. It is going to help you to maintain the independence of your functional requirements.

Axiom 2: This is the information axiom. This is going to help you to minimize the informative content of the design.

Cost-benefit analysis

Cost-benefit analysis, or CBA, is an approach that is meant to estimate the strengths or weaknesses of varies alternatives. It can be used with project investments, processes, activities, and even transactions. It can be used to determine, out of several solutions, which options will provide the best approach to a business in order to achieve benefits while still saving the company money.

To keep it simple, the CBA method is going to come with two main purposes. These purposes are:

- To determine if a decision or an investment for a business is sound. This means that the benefits will outweigh the cost. You also want to look at how much this is. If the benefits do not outweigh the costs much, then it is probably not the best option to go with.

- To help provide a good way to compare projects. This can involve comparing the total amount that you expect each option to cost against the benefits you expect to get.

The benefits, as well as the costs, are going to be shown in monetary terms, which makes it work well for Six Sigma. Moreover, they can be adjusted in the formula for the time value of money. This ensures that all flows of costs and those from benefits over time are expressed with a common basis.

The simple steps that you will follow when you are working on a cost-benefit analysis include:

- You first have to define the goals and the objectives of the project or the activity.

- You can list the alternative programs or projects that you may be able to use.

- List the stakeholders

- You then select the measurements you want to use in order to measure all of the elements when it comes to benefits and costs.

- You can also work on predicting the outcome of the benefits and the cost of each alternative over a period of your choosing.

- You can then convert all of the benefits and costs into a common currency to help them compare better.

- Make sure to apply any discount rates

- Next, you can calculate the net present value of all project options.

- Perform a sensitivity analysis: This is going to be the study of how the uncertainty of the output from a mathematical system can be shared to different sources of uncertainty in its inputs.

- After you have all this information, you can then pick out the option that is the best.

Root cause analysis

A root cause analysis, or RCA, is going to be a method to help with solving problems and it focuses on finding the root causes of the problem. A factor will be considered the root cause if you can remove it and the problem does not recur. Essentially, there are going to be four principles that come with this type of method including:

- It is going to define and describe properly the problem or event.

- Establish a timeline from the normal situation until the final failure or crisis occurs

- Distinguishes between the casual factor or the root causes

- Once it is implemented, and the execution is constant, the RCA is transformed into a method of problem prediction.

The main use of the RCA is to identify and then correct the root causes of an event, rather than just trying to address a symptomatic result. An example of this is when some students receive a bad grade on a test. After a quick investigation, it was found that those who took the test at the end of the day ended up with the lower scores.

More investigation found that later in the day, these students had less ability to stay focused. In addition, this lack of focus is from them being hungry. So, after looking at the root cause and finding it was hunger, it was fixed by moving the testing time to right after lunch.

Notice that the root causes are often going to come in at many levels and that the level for the root is only going to be where the current investigator leaves it. Nevertheless, this is a good way to figure out why one particular process in the business is not working the way that you want and then finding the best solution to fix it.

SIPOC analysis

If you are talking about process improvement, a SIPOC is there to be a tool that can summarize the inputs and then the outputs of at least one process and then shows it in table form. This acronym stands for suppliers, inputs, processes, outputs, and customers and these will be used to form the columns on your table.

Sometimes the acronym is going to be turned around in order to put customers first, but either way, it is going to be used in the same way. SIPOC is presented at the beginning of a process improvement efforts or it can be used during what is known as the define phase of the DMAIC process. There are three typical uses of this depending on who is going to use it, including:

- To help those who are not familiar with a particular process a high-level overview.
- To help those who had some familiarity with the process, but may be out of date with the changes in the process or those who haven't used it in a long time.
- To help those who are trying to define a new process.

There are also some aspects that come with this method that are not always apparent. These include:

- The customers and the suppliers are sometimes external or internal to the organization that is trying to perform the process.

- Outputs and inputs can include things such as information, services, and materials.

- The focus of this method is to capture the set of inputs as well as outputs, rather than worrying about all the individual steps that are in the process.

Value stream mapping

When it comes to value streaming mapping, we are talking about a method that is there to analyze the current state of a business and then designing a new state to use in the future. It is meant to take a service or product that a company offers from its very beginnings all the way through to when it reaches the customers. The hope is that the process is used to help reduce lean wastes, especially when compared to the process that the business is using right now.

The value stream is going to learn how to focus on any areas of a business that helps to add in value to the service or product. The purpose of this is to learn where the waste is in the business and then remove or at least reduce it. This can increase the efficiency of the business and can even increase productivity.

The main part of this process is to work on identifying waste in the business. Some of the most common types of waste include:

- Faster than necessary pace: This is when the company tries to produce too much of their product that it can damage the flow of production, the quality of the product, and the productivity of the workers.
- Waiting: This is a time when the goods are not being worked on or transported.
- Conveyance: This process is used to move the products around. It can look at things like excessive movement and double handling.
- Excess stock: This is when there is an overabundance of inventory. This can add on storage costs and can make it more difficult to identify problems.
- Unnecessary motion: This waste means employees are using too much energy to pick up and move items.
- Correction of mistakes: The cost that the business will have when they try to correct a defect.

This process is often used in lean environments to help look at and design flows for the system level. This is often something that is associated with manufacturing, but it can be used in many other industries including healthcare, product development, and even software development.

Business Process Mapping

The idea of business process mapping is going to be all activities that are involved when you try to define what a business does, who is the person or persons responsible, and at what standard a process in the business needs to be completed. It can also determine how the success of the process in the business can be measured.

Business process mapping is there to help a business become more effective. A clear business process map will allow even outside firms, such as consultants, to come in and look to see where improvements can be made, such as what can happen with Six Sigma, to help the business.

This mapping is going to take a specific objective of a business and they can measure and compare it to the objectives of the company. This makes sure that all processes that are done can align with what the company holds as its capabilities and values.

A good way to do business process mapping is with a flow chart. This can help you to see how the business does a certain process and can even include who is responsible for each part if that is important.

These are just a few of the options that you can choose from when it comes to working with Six Sigma. All of the options above can help you to make informed decisions while finding the process that is causing your business the most trouble at the time. Pick one of these options that go along with your biggest issue and find out how you can make smart decisions that will turn your business into something even better.

There are many consultants and courses out there selling "Lean" and "Six Sigma" and they can make the whole process sound like you need a Masters Degree to be able to grasp them.

In the reality this couldn't be further from the truth - the principles of Lean Manufacturing, Six Sigma and "World Class Manufacturing" are really very straightforward.

If you wanted to repair a car or fit a new door – you would need a toolbox. In your toolbox would be a hammer, spanners, screwdrivers, Allen keys, and perhaps chisels.

If you want to improve efficiency and reduce waste you will need a Lean Six Sigma "toolbox" – but instead of hammers and spanners you would have "5S", the "7 Wastes", "Changeover Reduction" and many more.

These are simple tools that can to be used for improving your processes, teams, business, and self.

If you keep this uppermost in your mind you won't go far wrong – it's not like you're going to have to try to learn a foreign language or master complex mathematics.

Use the particular tools that you think will apply to **your** process, for example "The 7 Wastes", and don't use the ones that don't, e.g. "Kanbans".

And don't feel daunted if you never use a particular technique – it may just not suit the business, the projects or you.

Chapter 7: Benefits

Lean, Six Sigma, Lean Six Sigma - these words seem to be tossed around like there's no tomorrow.

We have dedicated an entire book to teaching you the absolute basics of Lean Six Sigma and the ways it can be implemented into any business (regardless of its main industry).

What is the fuss all about, though? What are some of the things that make Lean Six Sigma such a success?

What are the reasons that make people turn to this method again and again and again?

Well, as it happens, Lean Six Sigma has been proved to be advantageous from a number of points of view.

In this chapter, we will elaborate on some of the most important benefits of embracing Lean Six Sigma - just enough to encourage you to proceed on this path!

Operational costs will be lower.

It doesn't matter what type of business you work for: operational costs are always considered to be significant risk factors. A high cost of operation means that you will make less profit - or that you might even lose money. And while we definitely agree with the fact that money is not everything in any way, it is quite important to acknowledge that money does make a business survive.

The Lean Six Sigma methodology can give you the road map you need to ultimately reduce your operational costs. For instance, if your lead times are very high, you will most likely find that you need to reduce them - and Lean Six Sigma can help with that by identifying the flaw in the process and providing you with a blueprint on how you can implement the necessary changes in this respect.

Cost Reduction

Your Lean Six Sigma approach can also help you reduce costs in your company, on multiple verticals. In general, all costs associated with inefficient and ineffective products or processes can be significantly reduced with the aid of Lean Six Sigma.

Some of the costs you can reduce by deploying and employing this methodology include the following:

• Inspection. Hiring entire teams of people to check your product is not always profitable, because most customers don't want to actually pay for this extra service. Lean Six Sigma can help you reduce the time and the costs connected to inspection (quality checks, assessments, etc.). When a good process is set in place, all potential errors will be captured in due time, and they will be addressed before they become actual problems.

• Rework. Clearly, having to re-do products or services for your customers cannot be beneficial to your business in any way. With Lean Six Sigma, however, you will have a methodology set in place to spot process errors before they become too big to handle and push your products into rework.

• Bad customer experience. Although bad customer complaints don't make your business lose money on the spot (unless customers ask for refunds), they can definitely affect your business in the long run. Therefore, it is quite important that you avoid them - and Lean Six Sigma can help you with that by pushing you to constantly improve the processes behind your products, according to the latest feedback received from the customer.

Boost Efficiency

Another major advantage of using Lean Six Sigma is connected to the fact that process improvement will eventually help you boost the efficiency of your business as well. Therefore, you will be able to generate more revenue in the same amount of time, using the same number of people.

Boost Accuracy

Lean Six Sigma is all about perfecting processes to the point they are spotless. Therefore, this methodology will definitely help you create more accuracy in terms of the data your business generates and how you can use that to avoid further errors in processes.

In Lean Six Sigma, the process is stable when it is in a state of statistical control. Therefore, this methodology can help you achieve the perfect balance between statistical accuracy and actual efficiency.

Furthermore, the same method can also help you comply with various types of regulations - both internal and external as well. This can be extremely important when it comes to the trustworthiness of your company!

Increase the Cash Flow

Cash flow is important for every type of business under the Sun - so it is important that you always have proper cash flow in your company. While Lean Six Sigma may not directly influence this, it will definitely help with the Days Sales Outstanding processes. Basically, the DSO should be as high as possible - so you can use goals related to this when setting up a Lean Six Sigma project. For instance, you could aim for invoicing process variation time reduction, for inventory management process variation reduction, and so on.

Beyond all the business benefits (quite numerous and quite amazing, maybe even more than we have managed to explain in this chapter), Lean Six Sigma is the kind of project management approach that will make you a better person.

Why does this matter?

Because great PMs are great people. They are people who lead by the power of example. They are people who set high goals in their personal lives - and most likely achieve those goals as well.

Great project managers are people who can organize their bookshelves just as well as they can organize their finances in a way that will allow them to buy a new car when they actually want it.

Sure, we can't promise that Lean Six Sigma will help you buy a new car - but what it can do, without any trace of doubt, is to help you aim for more and achieve more as well!

Benefits of Six Sigma

There are some companies that assume that they do not need Six Sigma because their business is doing well. However, they fail to realize the immense potential of the Six Sigma approach.

Those companies that have applied these methodologies and techniques have experienced substantial positive changes and immense benefits.

Here are some of the major benefits of Six Sigma:

1. Improvement in quality – Earlier in this chapter, we defined a defect as anything that fails to meet the expectations of a customer. Though these

expectations vary from one product/service to the next, it isn't that difficult for a company to know when one of its products or services has a defect. Six Sigma enables companies to reduce these defects as much as possible so that overall quality is improved. When this happens, more customers will enjoy the product or service.

2. Expanded innovation – If an organization wants to grow steadily and avoid stagnation, it must innovate. When Six Sigma came onto the scene, most companies believed that it would stifle innovation. However, the reverse has proven to be the case. Studies show that Six Sigma has promoted collaboration and new ideas in the companies that have embraced this methodology. Research shows that 46% of companies that implement Six Sigma see an improvement in innovation. One reason behind this is that the employees began focusing more on solutions instead of limitations.

3. Cost reduction – Six Sigma can save a company a lot of money by minimizing wastage and defects. The money saved can then be channeled elsewhere. For example, the US Army used the methodology in 2007 and managed to save a whopping $2 billion in that year alone. The army used Six Sigma to streamline some of its internal processes such as

task management, fuel recycling, and dining hall scheduling.

4. Focus on a common goal – In most organizations, getting every member of staff to focus on the same goal can be a tall order. The different departments tend to have their own individual objectives, and the only thing they share is the need to deliver products, services, or information to the customers. Six Sigma, however, brings the entire organization together to focus on the single goal of achieving a near-perfect performance level. The performance standard set by Six Sigma is 99.9997 percent. This is extremely high, especially if you compare it with the 99 percent performance standard that many organizations normally aim for. For example, if a company manufactures 1000 trucks, a 99 percent performance standard would mean that 10 trucks would be defective. However, using the Six Sigma performance standard of 99.9997 percent, only 1 truck would have a defect!

5. Continuous learning – An organization that implements Six Sigma methodologies will be in a state of continuous learning. The employees will have access to the tools necessary for generating fresh ideas that turbocharge development and innovation. When employees are relocated from one

sector of the organization to another, they bring with them a fresh perspective that will ultimately have a positive impact.

6. A boost in long-term revenue – When a company improves the quality of its products and services, it will experience an increase in its long-term revenue.

7. Enhances employee safety –Six Sigma always leads to a significant improvement in consistency and protocol within an organization. This is good for employee safety because employee working conditions get better. In fact, studies show that 56 percent of organizations that implement Six Sigma report an increase in employee safety. For example, a hospital that adopts the Six Sigma approach will be able to streamline laboratory processes. This will reduce the amount of lab staff overtime, thus saving the hospital a lot of money while also allowing staff to go home on time and rest. Employees who are well-rested are less likely to make mistakes on the job.

As you have learned, Six Sigma is an effective and efficient approach that focuses on improving quality, reducing costs, and increasing revenues. In the next chapter, you will learn the main tools and processes used in Six Sigma.

Six Sigma Tools and Processes

There are quite a number of tools and processes that are used within the Six Sigma methodology. In this chapter, you will learn what these tools are and how they work. You will also learn about some of the main frameworks that define Six Sigma.

The Benefits Associated with a Six Sigma Black Belt

Now that we have looked at all that the Six Sigma Black Belt can do for a person, it is time to look at the numerous additional benefits this type of certification is going to offer. There are benefits to the individual, as well as to the business that employs them.

- **Benefits to the Business**

A business having their own Six Sigma Black Belt program could be one of the smartest decisions that they ever made. This is going to allow them to train their employees in a way that will ensure that they are better for the company. Some of the benefits to a company that utilizes those with a Six Sigma Black Belt are:

1. The company can rest assured that they have the best working for them, and thus know that they are doing the best that is possible.

2. Customers are more satisfied with those businesses that have a person who is Six Sigma Black Belt, as these people are listening to customers and are going to put these comments to work for the business. Overall, it makes the business more customer-friendly.

3. More customers are loyal to businesses who have a person like this in their company. And more customer loyalty means a safer future.

4. Your bottom line is often improved when these professionals are on your team, as they are using whatever data they can in order to improve the overall functioning of the company.

5. Employees are happier when they have a line of work to follow that is going to render results without meaning that they have to do tons of things that are not helpful. Happier employees make for an entirely happier business in general.

6. Having these types of professionals on your team can also help make your business look better to other businesses out there. There are many businesses that are interested in partnerships, and this is one route to ensure that you can get these.

As you can see, a business can benefit greatly. Thus, it is advised that every business consider using these professionals or even having their own Six Sigma program in place to help train those who work for them. Otherwise, you cannot be certain that your business will thrive in the future.

- **The Benefits to the Individual**

Those who decide to take on the Six Sigma program and do become a Black Belt will find that they are going to reap many benefits as well. These benefits include:

1. The person will feel as though they are better at their job, and that is because they are. They are performing their job at a higher level than those who do not have this Black Belt certification.

2. They know that they have growth opportunities in the world of business because they hold a certification that makes them someone that businesses want to have on hand.

3. The methods that are used throughout this program can be used in other aspects of their life as well, meaning that they often gain something personal from this knowledge.

4. The job can become easier when they are using the right methods for obtaining information and analyzing the solutions that they are thinking of. This is something that many people do not think

about, but it is there once the person starts using their knowledge in their job position.

Overall, an individual is going to find that this can make them even better in the professional world. For businesses, there is no better way of improving upon their business than with the use of professionals who are Six Sigma Black Belt certified.

Personal Benefits of Lean Six Sigma

Following the discussion about the organizational benefits of Lean Six Sigma, we will now be discussing the personal benefits of Lean Six Sigma.

Lean Six Sigma offers benefits for individuals within the organization who finally results in being Lean Six Sigma heads. Outlined below are a few of the personal benefits you can get when you join the Lean Six Sigma project.

Personal effectiveness

Lean Six Sigma offers a well-detailed problem-solving methodology that could be used to look into any problem. Being able to discover and correct problems will increase your positive ability to work in any position and organization. The Lean Six Sigma leads you through a structured process of inquiry, problem identification, analysis, and solution formation. A lot of the techniques and tools can be used for your problems every day. However, if you use only a few techniques and tools, the structured problem-solving approach will put you in charge of locating and correcting your problems.

I have made use of this approach when correcting and solving problems in my home, with local charities I back, and in multiple businesses.

Leadership opportunity

Lean Six Sigma, effective through projects, is headed by leaders.

Leading a Lean Six Sigma project will regularly offer you the chance to know other functions as well as senior management. The exposure is in the area of an individual who can locate and correct a problem.

Communicating with team members and managers will, in most cases, increase your level of communication and decision-making abilities. The organization of the Lean Six Sigma can assist you in building your project management skills. Having the chance to put on your CV that you headed a project team that obtained quality improvement, lesser cycle time, and cost savings will aid you as you are searching for that subsequent promotion or new chance.

Pay and Promo ability

This is the last personal benefit of Lean Six Sigma. Achieving belt certification is a vital record of your CV. A lot of job postings need an applicant who possesses the necessary and required Lean Six Sigma credential. In this way, you stand a chance to get promoted.

Also, within an organization, promotions are always on the basis of how you have shown your leadership skills to the team members. When you lead a Lean Six Sigma team successfully, it will demonstrate senior management and HR that you have gathered for some time and which will also prep you for higher duties.

The average yearly income in the USA for Lean Six Sigma Black Belts is about $100,000. Meanwhile, the average yearly income will also depend on your industry and the country. Finally, it is without doubt and argument that the Lean Six Sigma certificate will increase your chances of earning high.

Chapter 8: Criticism

- **Limitations and criticisms**

    Six Sigma is often seen as a revolutionary and powerful management tool thanks to the performances recorded by the many companies that have adopted it. However, like all methods, it does have some limits, both methodologically and terminologically. Furthermore, as is the case for many other economic aspects, there is a difference between the theoretical and practical aspects. American economist George Eckes, Six Sigma specialist, highlights the failures often observed during applications of the method and offers some recommendations:

- **Consider that quality improvement does not result only from improving statistics.** Rigour and discipline can be significant assets, but they do not cover all the means necessary for the proper management and improvement of a process. Six Sigma combines a series of complementary areas and does not neglect the human aspect in any case, which is both an actor (employees within the company) and a target (customers to satisfy). This aspect is often overlooked during applications within a business.is

- **Realise that reducing costs is only one step of the improvement process.** Six Sigma does not consist of programming cost reductions for strategic purposes. On the contrary, this method advocates efficiency and effectiveness by refocusing the company objectives on customer expectations, rather than an accounting approach that calculates the known costs and neglects the impact on the customer.

- **Be sure to include improvement in job descriptions.** It is not always easy to reform a process in a company in order to apply Six Sigma. Employees or workers often feel they don't have time for such a re-assessment and believe that they already devote sufficient time to the company. Yet this 'surplus' of time they spend on working for the company is often due to ineffectiveness and inefficiency. This does not necessarily come from the unwillingness of the worker, but rather the process itself.

- **Remember that team dynamic is a leading cause of project failure.** Although it seems easy to manage team dynamics, this is one of the main sources of failure. It is therefore important to build a solid foundation. To do this, the project manager must clearly explain the ins and outs of the project. Meeting supervision, setting the agenda and determining the respective roles and responsibilities are starting points to ensure that the project does not begin on shaky ground.

- **Consider that the Black Belts are not completely responsible for the efforts.** Black Belts are intended to be team leaders. As explained above, they are usually people trained in the use of tools and techniques for improvement – almost like operational leaders. The danger lies in the fact that everyone (including the leaders of the company) separates themselves from the responsibilities of the project as they imagine that domestic experts are there to launch Six Sigma. However, the proper functioning of a company comes from teamwork, and all hierarchical management positions are involved.

- **Consider Six Sigma to be an improvement in continuity.** One of the principles of the method is to work in continuity and constantly ensure a quality process, not to form a team in charge of Six Sigma as soon as an inefficiency or ineffectiveness problem arises in the company.

- **Think of management as an active player.** For Six Sigma to work, the leaders of the company must get their hands dirty and consider themselves participants in the work of the company. Senior management is aware that cultural phenomenon is an important element in business management. One of Six Sigma's strengths is that it encourages a proactive attitude at all hierarchical levels.

- **Be aware of the changes in business management.** If changes at strategic levels are not well managed by the company, the potential results will remain low.

- **Related models and extensions**

Lean Six Sigma (LSS)

Lean Six Sigma (LSS) is an extension of Six Sigma that is becoming increasingly important. It is more focused on the production process, while Six Sigma focuses mainly on the product itself. This related model allows you to reduce the working time and waiting periods required for the establishment of a more effective process.

The strategic objectives of this model are:

- increasing the added value of process tasks;
- reducing the time and cost of the process by eliminating activities with no added value;

- making processes more fluid;

- improving the quality of products according to customers;

- encouraging the development of a culture of continuous improvement within the company.

- The main areas of action are:

- defining value and identifying the steps that create it;

- identification and elimination of waste and hidden costs;

- control of variation sources by following the steps of the process.

Total Quality Management (TQM)

Total Quality Management is an older quality management approach than Six Sigma. Their common objective is to mobilize the whole company to achieve perfect quality while reducing waste and improving the final product through performance. TQM focuses on the customer – satisfaction and loyalty – although the practice of quality control and self-control is essential here.

The methodology of the model is as follows:

- **Plan.** Development of strategic objectives and schedule improvement plans.

- **Do.** Implementation and application of improved production processes.

- **Check.** Satisfaction analysis and quality control of the product.

- **Act.** Correction of costs and waste and control of production stages.

According to US project management Frank Anbari, Six Sigma is more complete and comprehensive than TQM because it provides financial results and combines advanced analysis tools and managerial methods. It also summarises the relationship between the two methodologies: Six Sigma = TQM + customer focus + complementary data analysis tools + financial results + project management.

Common Six Sigma Critiques

One common reason that Six Sigma implementation fails, is that many of those in management positions have never heard of the system and as such have a number of off the cuff questions or preconceived notions about it. Forewarned is forearmed however, so many of the common critiques of Six Sigma, and their rebuttals, are listed below:

Six Sigma is just a fad like any other flavor of the month management style.

In all actuality, Six Sigma can trace its origins back to the early 1900s where it was pioneered by the likes of Walter Shewhart,

Henry Ford and Edward Deming. What's more, it separates itself from the pack of related programs dedicated to continual improvement by being more focused on the use of data to make appropriate decisions that focus on the customer and ultimately always provide a solid return on any given investment.

We don't have the resources or time to dedicate to teaching and learning Six Sigma

Time is undeniably the most important resource that any company has, as it is the only resource that is truly finite. Instead of considering the amount of time and resources that investing in Six Sigma will require, a better metric would be to determine what not adhering to a more effective system such as Six Sigma will cost in the long run.

It is important to remember the story of the pair of lumberjacks who worked day after day in the forest. One man worked himself to the point of exhaustion every day while the other man spent the time preparing properly, and at the end of the day both men had always chopped the same amount of wood. If your team has the opportunity to work smarter instead of harder, why wouldn't you provide them with the tools they need to make that the new norm.

What's more, the cost of training the team in the specifics of Six Sigma can be mitigated over time by spreading out training courses as needed. While you won't start seeing the benefits from the higher levels all at once, training the whole team to

Yellow Belt certification will still produce noticeable results. In addition, any funds put towards this type of training can also be seen as investing in the future of the business and should be considered accordingly.

We're too small for Six Sigma to be effective

Six Sigma offers up a new way of looking at day to day business interactions that will increase productivity, and ultimately profits, regardless of the size of the team in question. Will a 10-person team need as many training sessions and resources committed to the project as a 50-person team? Of course not, but the individual results will be the same.

Six Sigma doesn't apply to us

While Six Sigma originated in the manufacturing sector, studies show that industries based around providing services are actually more likely to generate unnecessary waste, not less. This is caused by the fact that so much of what is provided is essentially intangible which makes standardizing processes much more difficult. However, all of the processes that are already in place to track the services being provided can ultimately be leveraged to implement Six Sigma successfully.

Six Sigma involves too many statistics to be used practically

Despite its reputation for being all about the numbers, a majority of the tools and principles that are used in implementing Six Sigma require more common sense than mathematical formulas. For example, mitigating waste is one of the most important facets of Six Sigma, a facet, that only requires an understanding of the process in question and how to do it in the most effective way possible. Operating under Six Sigma is more about fostering a mentality that allows employees to get to the root cause of an issue, regardless of how long it takes. The formulas and mathematical equations simply justify it after the fact.

Lean is a better fit for us right now

Lean and Six Sigma aren't opposing processes, and indeed they work incredibly well together, so much that Six Sigma is often referred to as Lean Six Sigma. When used in conjunction with one another, Lean will improve the throughput and speed of your processes while simplifying and allowing the team to do the best with what's have available. Six Sigma then takes those improved processes and makes them as high of quality as possible by reducing deviation and related defects. Combing the two will only lead to better results overall. Starting with 5S is recommended, after that there is no harm in combing Lean and Six Sigma for the best results.

We've tried it before and it didn't work

The Six Sigma system has a proven track record with some of the biggest corporations in the world, including Starbucks and Coca Cola. This means that the reason for the failure is likely related more to the way the program was managed in the past as opposed to a lack of efficacy when it comes to Six Sigma in general. Looking back, it will likely become clear that there was previously a lack of clearly defined goals or possibly unreasonable timetables applied to its implementation. Given the right allocation of resources, there is no reason that any company cannot achieve success with this system. As long as the goals for success or failure are clearly defined, there is no reason that the success from this attempt should not completely overshadow any previously failed initiatives.

Deciding if Six Sigma is Right for Your Company

While the Six Sigma system has something to offer teams and companies of all sizes and complexities, that doesn't mean that it is automatically a good fit for your particular business. In fact, implementing Six Sigma successful depends on a myriad of different specifics, including the conviction of those implementing the system, and the company's overarching culture. This is why it is suggested that you start by implementing 5S as discussed as a way to ease your team into the idea of Lean systems. As such, understanding the concepts related to successfully implementing 5S will make the transition to Six Sigma much easier to accomplish.

Is the leadership involved?

When suggesting a transition to Six Sigma, it is important that it not be framed as another "fad" management style and instead be seen as an enhancement of what is already in place. As a rule, management is going to be opposed to the change, which means it will be important to get buy-in from the person at the top and work down from there. It is important that the company culture is one that supports positive change and remember, if the management team can't be seen as achieving a consensus on the new program, it is already dead in the water.

This doesn't mean that every person in the organization must be committed to Six Sigma, but it does mean that the change must be seen as institutional, which means the public front must always appear united. The human brain is a creature of habit, especially when it is confronted with new systems that seem complicated which Six Sigma often does, if it is at all apparent that the new system is optional, most people will opt out every time. Don't give them an avoidance opportunity, do what you can to ensure that opposition is voiced in private.

Is the correct infrastructure in place?

Six Sigma is founded on the principle of leaders mentoring those underneath them, and in order to make Six Sigma work, this needs to be a full-time job for some people, at least until new, positive habits form. While it may not seem cost-effective to dedicate one or more people to the task of actively mentoring others on Six Sigma specifics, it is a sacrifice you must be willing to make if you want Six Sigma to be more than a flash in the pan with your team.

Unfortunately, this will never be the case if the person who is responsible for mentoring others is also bogged down with additional work as that additional work is almost always going to ultimately end up in front of additional mentoring duties. Prior to implementing Six Sigma it is important to ensure that you have the infrastructure to support it long term, otherwise you are ultimately just wasting everyone's time.

What will cause the rank and file to fall in line?

Once you have the support of management and have ensured that you have the infrastructure to support the undertaking for as long as it takes, the next thing you need to determine is if you have a way to motivate the remaining employees to stick with Six Sigma to the point that they internalize it so that it becomes second nature. Regardless of how their progress is tracked, it is vital that each member of the team feels an immediate and compelling reason to commit to the new program, at least at first.

Companies are like any other body that is in motion, the larger the company, the more inertia it displays when it comes to making large changes, and for many companies Six Sigma is a very large change. This is why a tangible incentive must be attached at every level of the company to ensure that everyone remains united in their drive to obtain the incentive. The side effect from this, will of course be that they are also internalizing the ideas behind Six Sigma without actively trying to do so.

How common is the practice in your field?

While Six Sigma has proven value in a wide variety of fields, that doesn't mean that all of those fields are ready to adopt the process with open arms. While being a forward thinker is never a bad thing, if your industry has yet to adopt Six Sigma as a common practice, it is important to be ready for additional pushback from the institution when attempting to move forward with the change.

Be prepared for resistance and stand your ground, using examples of successful companies who have already adopted the Six Sigma system will also help to silence naysayers. The research behind Six Sigma speaks for itself, be ready with specific examples of how it can help your company specifically and the facts will ultimately speak for themselves.

What are the objectives for the training?

Depending on the size of your company, training the team at the same overall level of Six Sigma may make sense. Eventually, however, the size of the team will necessitate the use of numerous training levels. If this is the case, it is important to consider the qualifications for each level as well as how training will be staggered for maximum efficacy. Don't forget to determine how the length of the training will affect other duties as well as the areas that will be focused on the most.

Taking the time to identify the specifics unique to your desired training scenario before you start will make all the difference in the overall implementation of Six Sigma and should not be ignored. There are no one-size-fits-all options in this scenario, planning out the specifics of your team's training could very well make the difference between success and failure in the long term. What's more, it can help point out potential issues that may arise which may otherwise have not been visible until the training was already in progress.

Which projects are going to be used as the flagships for the new system?

Once Six Sigma training is completed, you will want to already have a few projects waiting in the wings that can be ultimately connected to the new system and pointed to as signs of success further down the road when the question as to whether or not it is a good idea to continue with Six Sigma arises. Your goal should be to increase the level of involvement surrounding one or more Black Belt or Green Belt projects and do whatever is required to ensure they are successful.

Chapter 9: Lean Six Sigma Certification

How to Get Certified?

Since Lean Six Sigma is such a popular project management methodology, you will most likely find a training program near you. Every type of certification starts with proper training, even if you have already undergone it for a different level of your Lean Six Sigma certification.

Enroll in the training program and get ready to feel like you're attending school all over again. You will have a lot of classes, you will learn a lot of things, and you will collaborate with other trainees just as if they were your old classmates.

It's going to be quite fun, actually!

Once your training is done, you will have to sit an exam and complete a "test project."

Given the fact that you already know a lot about Lean Six Sigma and that you will learn even more during your training, you will definitely take your exam and get your certification!

How to Get Six Sigma Certification

Six Sigma is a project management methodology that is there to help increase profits, ensure product quality, boost morale, and reduce defects. Many companies use it in order to help them strive to be as close to perfect as possible. Although there is not really a governing body that will dictate the rules of Six Sigma, many organizations are going to offer certification in this methodology. By becoming certified with Six Sigma, you will find that you are someone to take seriously and that could provide more value to the company you are already with. Some of the things that you need to do in order to get Six Sigma certification includes:

Determining the management philosophy

1. Consider what the organization needs: What kind of management style is going to benefit your organization the most? Is it dealing with too much waste or overhead in the supply chain? Are there some issues with staying consistent to get things done? What is the overall culture in the business?

2. Decide how you would like to optimize the process: You may be someone who thinks that the best way to make sure quality is there is to ensure that all the processes are consistent with as few variations as possible. Others may want to opt for efficiency or producing a quality product without as much waste as overhead.

3. Determine which certification you want to go with. You can go with Lean Six Sigma or Six Sigma. The

type of management philosophy you go with will determine the answer to this.

a. Six Sigma is going to define the waste as a variation in the process of the business. If you believe in a consistent process, then you should get this certification.

b. Lean Six Sigma is going to be a combination of the Lean method and the Six Sigma method. It is going to define waste as anything that does not end up adding value to your product when it is done. If you would like to be more efficient, then this is the best option to choose.

Decide the level of certification

1. Determine your role in an organization. This can determine how high of a certification you should get. Are you someone who supports the manager or the manager? Does your work involve you just being able to use Six Sigma on a project?

2. Consider your goals in the future: Even if you are not a project manager right now, if that is your future goal, then you should consider this when picking out your certification.

3. Select the certification: There are four levels that you can choose including Yellow belt, Green belt, Black belt, and Master Black belt:

    a. Yellow Belts: These are those who have just a basic understanding of the process. They are more of a supporting role to those with the higher belts. There aren't many courses for this because it is so basic and most concentrate on being an expert in the field.

    b. Green Belts: These are the individuals who will work closely with the Black Belts and are mainly responsible for collecting data. These individuals will use Six Sigma, but they often have other responsibilities outside this project, so they still just need a basic understanding.

    c. Black Belts: These are individuals who are project managers. The other individuals will report to them and they are going to be the ones who are going to dedicate a lot of time to the project.

    d. Master Black Belts: These will be the experts of the team. They will be the one that the team will turn to if there are any errors or if they need to make some corrections along the way.

Getting certified

1. Find a training program: It is likely that you will have to do some classroom instruction, so look and see if there are some near you to avoid travel. Always make sure that the program is accredited. No, there are no formal standards right now, but there are some accreditation organizations that can make sure you actually learn what you need.

2. Enroll in the program: You will attend the right classes and learn the material that you need to get the belt that you choose.

3. Take the written test: Once you are doing with the training, the next step is to do the written test. This will check to see if you have learned what you need about Six Sigma. These tests can take some time. The Yellow Belt can be two hours, the Green Belt about three hours, and the Black Belt about four hours.

4. Complete the projects: The final phase of being certified will be the process of completing a few projects using the Six Sigma methodology. This is like the "lab" to make sure that you are able to implement what you learn.

Benefits of being certified

Employees who are certified in Six Sigma can bring many benefits to a company. You already know how the process works and can be there to help the project go smoothly. It does not matter which belt you end up getting, they all are important to implementing a project and ensuring it is done.

If your current employer is looking at getting started with Six Sigma, it is definitely worth your time to get a certification. This can ensure that you are put on the project and can really help to further your career, especially if you see success. If your current employer is not getting started with Six Sigma, it is still worth your time. You may be able to use this later on with another company or whenever your current company decides to look as well.

That is all there is to it. The amount of time that you take to complete the certification is going to vary based on which belt you want to get and how the training centers work in your area. With some time and studying, you can learn how to make Six Sigma work for your business.

Chapter 10: 10 Six Sigma Do's and Don'ts

Any organization that wants to implement Six Sigma must be willing to follow a systematic approach. There are many organizations that merely use Six Sigma as a way to market and promote their products but are not really interested in applying it as a long-term business initiative. Such organizations have failed to realize the value and benefits of Six Sigma.

You will learn about some of the things that an organization should do and some of the mistakes that must be avoided. All throughout this book, you have slowly learned what to do and how to adopt a Six Sigma approach.

But you also need to keep an eye on the assumptions and useless approaches that some organizations make. These pitfalls tend to be disastrous for the organization, and a failed Six Sigma initiative can have terrible financial repercussions.

With that said, let's look at 10 do's and don'ts when deploying Six Sigma.

The 10 Do's of Six Sigma

1. Do explain why you need to adopt a Six Sigma approach – The first thing you must do is establish a

clear reason why Six Sigma is needed and how it will benefit the entire organization. This will help you convince not just the upper management but the lower level employees as well.

2. Do get the support and engagement of upper management – Make sure that top management is sold on the idea of improving the current processes. Once the organization's executive team understands the potential of Six Sigma, they will provide the impetus needed to make it a success.

3. Do ensure proper project planning – You have to think things through before taking any action. This is one of the benefits of using the DMAIC framework. It forces you to define, measure, and analyze the problem and come up with the best solution possible.

4. Do create a robust project selection system – The project team must start things off on the right foot. This means they have to ensure that the right project is selected based on reliable data and strategic business goals.

5. Do choose the best candidates for Six Sigma training – You cannot train the whole organization when deploying Six Sigma. Therefore, pick employees who have a keen interest in Six Sigma and train them so

that they can transfer that knowledge to other employees. With time, the majority of employees will have gained practical knowledge of the Six Sigma methodology.

6. Do select a capable Champion – Every Six Sigma initiative must be led by a Six Sigma Champion. This should be someone who is trained in Six Sigma and is passionate about its implementation within the organization. It will be the Champion's responsibility to get everyone else excited about the changes and benefits that Six Sigma will bring.

7. Do involve all relevant stakeholders – Implementing Six Sigma needs to be a collaborative effort if it is to succeed. Everyone, including the lowest level employees, customers, and suppliers should be involved in the implementation process. The process owner or sponsor should also be consulted in every phase of DMAIC.

8. Do ensure maximum communication – There must be effective communication at all levels of the organization. This will ensure that everyone stays in the loop and no department feels isolated or left out during implementation.

9. Do review the deployment regularly – Once the improvement project has been implemented, it must

be monitored regularly to ensure that everything stays on track. This will enable any deviations to be identified early and corrected.

10. Do celebrate your success – After you successfully implement that first Six Sigma project, make sure that you celebrate the victory. Success tends to be contagious, so when people recognize and praise the victory, greater interest in the Six Sigma approach will be developed.

The 10 Don'ts of Six Sigma

1. Don't let Six Sigma be a fad – Some organizations view Six Sigma as a way to appear in tune with current best practices, so they use a few tools for a while but don't fully adopt the methodology. Any organization that does this will never realize the full potential of Six Sigma.

2. Don't assume that you are different – Many organizations refuse to adopt Six Sigma because they believe that they won't get the same results as other organizations. While it is true that every organization is unique, Six Sigma can still work and benefit any business.

3. Don't hire part-time Black Belts – One common assumption is that Six Sigma Black Belts can work as

part-timers and the project will still proceed at a good pace. This is a mistake because, without a full-time Black Belt, the project will not generate the momentum required to bring change to the organization.

4. Don't implement without a deployment leader – The role of the deployment leader is to train the project team, assign tasks, and choose the Six Sigma tools. If there is no deployment leader, there will be confusion and conflict that will ultimately result in failure. The team members will begin to focus on their own areas and there will be a lack of genuine synergy.

5. Don't get greedy – One of the main reasons why a Six Sigma project ends in failure is scope creep. If you get greedy and try to fix every problem in the organization, you will fail. Avoid making the project scope too broad. Otherwise, the project team will easily be overwhelmed.

6. Don't focus on isolated implementation – Six Sigma is best used as a tool for improving an entire process rather than just a single product or service. Isolated implementation may be a worthwhile idea if an organization lacks adequate resources, but it is not a smart strategy in the long run. You cannot actualize

the full benefits of Six Sigma through small and disconnected improvement projects.

7. Don't obsess over Six Sigma training – The common assumption is that everyone involved in executing a project must be trained and certified in every tool and technique. This is not true. There are a lot of Six Sigma courses and complex tools that are being taught, but you don't need all of them to make your project a success. It is better to focus on learning and applying what is relevant to your project. There shouldn't be an obsession with the number of Belts in the project team.

8. Don't ignore technology – Technology plays a critical role in implementing a Six Sigma project. Technology makes it easier to measure, analyze and control systems and processes that need to be or have already been improved. It would be unwise to try to separate the two.

9. Don't forget to validate the measurement systems – Six Sigma is built on the foundation of hard data and precise measurements. However, don't fall into the trap of blindly following a measurement system without verifying it.

10. Exaggerating the opportunity counts – Sometimes Six Sigma practitioners try to exaggerate the number

of opportunities there are in a process. This is usually an attempt to fabricate their performance so that stakeholders will believe that the process has improved significantly. This kind of deception is wrong.

Chapter 11: Six Sigma Success

Is Six Sigma successful, and can you be successful as a certified yellow, green, or black belt? It's possible. There is a lot of scrutiny for the Six Sigma program. You've seen the foundation, that there is a methodology based in psychology and focusing on manufacturing processing. It's an odd combination, but it's not unrealistic. For centuries people have put the work and principles of psychology into motion in the workplace. That doesn't mean that Six Sigma is a 'quack' method of getting people to spend money although there are numerous people out there who believe that about Six Sigma.

In fact, there are entire organizations dedicated to exposing the failures of Six Sigma and Six Sigma teams. Is it a scam? No. It is a methodology, and with that will come both success and failure. Throughout this book, as you learned about various aspects of the philosophy, tools, and goals of Six Sigma, you probably became aware that there is no one-size-fits-all process that will lead every company to success. There are too many human elements and external factors present to create ultimate success again and again and again. But realistically looking at it, you have some outlandish claims that support Six Sigma. Such as "Honeywell recorded over $800 million in savings." Is that realistic? No. It's not practical, and these claims are similar to the get-rich-quick schemes that you cringe at on television shows and in commercials.

However, with clear set goals and expectations, you can have success with Six Sigma. Does every Six Sigma process aim to help the company save hundreds of millions of dollars? No, it doesn't. In fact, one Six Sigma initiative led by entomologist Don Messersmith only aimed to reduce the deterioration of the Lincoln Memorial. The confusion that Six Sigma is about saving companies millions or billions of dollars comes from Six Sigma organizations claiming that they can do it. The issue is that most companies don't have the opportunity available to save that much money. Super companies clearly have problems with waste management, process improvement, and defects on a much grander scale. Honeywell, GE, and AT&T have the perpetual business to make these claims of saving millions or billions. Small and medium-sized companies don't have the flux in their revenue or their work.

Returning to the Lincoln Memorial example, what was the solution to reduce the deterioration? Through the 5S tool, Messersmith was able to identify that the deterioration came from harsh chemicals used for cleaning bird droppings. The bird droppings were in such volume because there was a healthy population of spiders, which the bird's liked to eat. The spiders were there because there were numerous insects and the insects came in because of the lighting that switched on at dusk and off at dawn. The solution? Change the lighting. Much of the Six Sigma success comes from common sense or principles that we all know and have difficulty applying. Similar to the stance of the *7 Habits of Highly Effective People* from Stephen Covey. Six Sigma says here is a principle that applies all the time. Focus on reducing defects, creating consistent products, and then you will see success. Or from another master of the industrial revolution, Dale Carnegie voices in *How to Win Friends and Influence People*, "Action breeds confidence," and as mentioned earlier, Six Sigma is about taking action when otherwise companies would prefer to do nothing. That drive to do nothing or to take the easier route is the same thing that leads companies to believe that Six Sigma projects fail after the professional leaves the project. Companies often fail to maintain the program or to follow up on optimization efforts after the Six Sigma professional moves onto another project. The Six Sigma program has many safety nets in place to prevent this from happening, but the responsibility falls to the company. This problem is why many companies are embracing Six Sigma as a company-wide method of operations. Rather than choosing to use Six Sigma for standalone projects when the occasional problem crops up. Within the companies that fail to maintain a system, it's often clear from an outside perspective what happened. Someone with the company, specifically within mid or high-level management, either didn't care enough about the system or didn't understand the underlying concepts of Six Sigma.

Ultimately as a Six Sigma professional, you will spend your time working with many concepts, which business professionals know but fail to put into motion. The position of compelling business professionals to do the right thing for their company can become frustrating and at times, tiring. You will need to create motion and action where there is stasis. You will need to cultivate consistency where there is complacency. Six Sigma teaches you the philosophy and how to use particular tools that will put you on the right track. However, many people go through the training and feel that they didn't get much out of it because they still struggle with the base level concepts. What are the career possibilities for you if you do choose to go through with Six Sigma training?

On a yellow belt level, you can expect to pursue careers in supervisory and management capacities. Yellow belt training encompasses a lot of data collection and builds people skills for management level interactions. That makes it reasonable for yellow belts to work as Sr. project managers, association project managers, service managers, project coordinators, and business specialists. However, they can often find work specifically as a Six Sigma analyst or similar job title. As a yellow belt, you will likely put in about half of your time at work on Six Sigma related initiatives. However, this level isn't quite ready to work on these projects full-time. Yellow belts are well-known for having prowess in learning systems and processes quickly, making them well-suited for changing from one project or department to another. Not to mention that many people who pursue Six Sigma do so because they don't want to sit behind a desk and do the same thing every day. Working in different fields and improving processes are usually pretty high on their goals. If this sounds good to you, then know that you may only spend a short time in these positions while you work on your green belt, which has many other career opportunities available.

Green belt certifications open up a lot of doors for Six Sigma professionals. Not only can you go into full-time Six Sigma work, but you may need to. In order to continue growing and achieve your black belt and mastery black belt, you need to complete at least two successful Six Sigma projects. Seeing projects through to success could take weeks, months, or years. Additionally, you need to begin working as a mentor, which can put a lot of strain on your time management as well. Green belts will often work as consultants, process engineers, and manufacturing professionals. Many engineers, such as automotive and product engineers, find it imperative to obtain a green belt to reach mid-level management positions. For those who don't have a strong engineering background, you can pursue employed as a specialist, process developer, data scientist, and business process analyst. However, to reach chief-level positions, or upper management, you may need to continue on to your black belt. Most black belts work as high-level managers or independent consultants. You might want to explore careers in quality assurance management, project directing, consulting or operations directing. Many of these options fit well with the tools and skill you build through years of Six Sigma practice.

It is worth noting that there are very few career paths, which require a Six Sigma certification. There are, however, many jobs, which explicitly require the certification. Companies that want Six Sigma executed internally will employ people based entirely on this certification alone. Give these certifications the same level of thought that you would give to the university education. Can people succeed as Chief Executives without a degree in anything? Yes, they can, and many do. Famously Larry Ellison, Bills Gates, Steve Jobs, and Mark Zuckerberg are all college dropouts. That doesn't make them any less adept. However, it does make them more reliant on the circle of people around them. They may opt for more experienced people who do have proof of their skillset. Additionally, many of these people are exceptions to the common rule that educated tend to obtain higher levels of employment. Most often, you can expect a formal education to give you some advantage in the workforce. That doesn't mean that you don't' need real-world experience to quantify your education and skillset. When working as a Six Sigma professional or looking for work as a Six Sigma professional, you should craft your resume to showcase both. In addition to your work experience, list your experience exclusively on Six Sigma initiatives. Even if they were within an employer, you could show that these projects stand apart from your work experience. Be honest as well: which projects were a success and which weren't. Many companies that are looking to hire Six Sigma professionals already know that not every project is a success. However, you can work diligently to create success wherever possible.

There are a few ways that you can focus on the mission-critical goals of Six Sigma projects and build a better chance at success. First, always seek out the root cause. You'll learn this in white belt training, and it's vital throughout every Six Sigma project. Second, make sure that you're using the right tools. When working as a yellow or green belt, you might feel inclined to stick with the tools you know. However, the success of any project depends largely on using the right tools. As you go through your Six Sigma training, you'll have to work and learn to go outside of your comfort zone. This trait is something that companies expect when hiring a Six Sigma professional. They want someone who can work with new or innovative methods and technologies with confidence. Finally, you must always refer back to your data. Perhaps the most common job for green belt Six Sigma is working as a data scientist. Really for any green belt position though, you need to have a strong grasp and confident command of data handling and data-driven decision making.

When looking at Six Sigma as a whole, and whether or not you decide to pursue formal training, consider the skills and techniques that you will learn. Of course, there is no guarantee of employment. University degrees don't come with a guarantee of finding a job after graduation. However, through the Six Sigma programs, you'll learn invaluable tools that you can put into effect in nearly any position — aiming for zero defects and understanding that you need data to drive decision making. One of the most powerful skills you might learn is how to obtain a buy-in from high-level management. If you want to cultivate change and make waves in your industry, then you will need these tools. For many people, that's enough to want to begin training and start pursuing a career path that fits their ideals best.

Six Sigma is many things. It's a methodology for business management, which contains many tools for increasing quality and helping build a platform for business success. Driving the Six Sigma methodology is a philosophy of creating the best and most consistent products or services possible. These goals are exceptional and amiable. They're also goals that benefit every possible industry. As companies are looking for cutting edge talent, it's not shocking that they're looking towards certifications that benefit many departments or divisions rather than a single skill. With Six Sigma and practical experience you can become successful in management and leadership roles. Many Six Sigma professionals work as successful independent consultants. While other Six Sigma professionals create processes and design products using the skills they learned through certification. Success is out there, but one certification won't guarantee it.

The Six Sigma methodology provides a great platform for a successful execution of performance excellence and continuous improvement projects. However, there are some factors that can also affect the deployment of Six Sigma and the completion of the projects. These include company and project team structure, communication, customer insight, and stakeholder management.

Many businesses exist as a Hierarchical structure. The risk associated with this is that departments drive their own agendas and sometimes fail to consider the bigger picture or overall objectives of the business. A key to successful Six Sigma project completion is designing a team that will deliver the desired outcome in the most efficient way. Usually, this involves the formation of a cross-functional, multi-disciplined team. A very good trick employed by experienced Green Belt Project Leaders is to use or identify Champions or Role Models within the business. By nature, these people tend to be highly driven, committed and change agents within the existing business. Many will have a good communication style and will have the respect of the existing staff. Also, a project is only successful if the improvements are sustainable. Selecting a team that will not only drive improvements but also create a culture for success is critical for any Six Sigma project.

Support of top management and good stakeholder management are essential for success in Six Sigma. Whilst it is true that many executives and top management will be somewhat removed from direct processes, they will directly control the overarching strategic direction of the business, budgets, training programs, succession planning, and cascading of messaging.

In his book "The Five Dysfunctions of a Team", Patrick Lencioni describes a model that outlines the inhibitors to building great teams.

At the base of the pyramid is an absence of TRUST. A trusting team will:

- Admit mistakes
- Ask for help
- Accept questions and inputs into their area of expertise
- Focus energy on issues, not politics
- Look forward to team meetings

Next is a fear of CONFLICT. Teams that engage in constructive conflict will:

- Have lively and interesting meetings
- Extract input from all team members

- Solve problems in an efficient manner
- Minimise politics
- Open critical topic for discussion

The next segment of the model is a lack of COMMITMENT. A committed team will:

- Create clarity around direction and priorities
- Align the team to common objectives
- Develop the ability to learn from mistakes
- Move forward without hesitation
- Change direction to gain competitive advantage

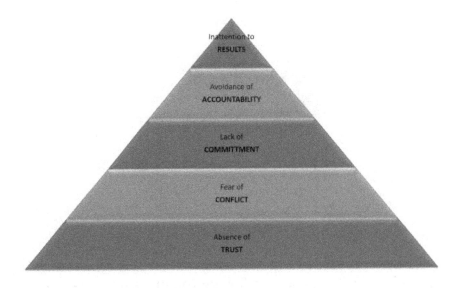

Five Dysfunctions of a Team

Throughout the different stages of the Six Sigma project, the Green Belt must learn to adapt their style of leadership, this is especially true when analyzing what stage the project team is at. When building a successful team, there are four stages that the team will transition through. These are: Forming, Storming, Norming and Performing. Successful teams all pass through these different stages and even the great teams need help and guidance along the way. It is very important that the Green Belt shows strong leadership when required in the early stages and learns how to guide without micro-managing once the team is starting to gel and execute tasks (Norming - Performing stages).

Stages of Forming a Team

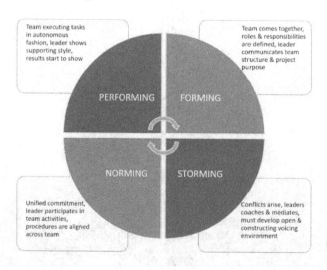

Successful Six Sigma comes for continuous improvement and striving for Performance Excellence. You do not need to be a master of statistical interpretation or data analysis, but rather a disciplined and persistent team player. Ultimately, successful Six Sigma projects are driven by teams that recognize their mistakes, identify the opportunities for improvement and focus on results

So finally, I must say well done, you have successfully started your journey to Performance Excellence. I truly hope you have enjoyed this introduction to Six Sigma and gained a better understanding of the role played by the Green Belt. The road ahead will be a challenging one but always remember there is a clear process and a defined set of tools to support you along the way. If you trust in this and the team around you then I have no doubt that you will be successful and achieve your goals - Brendan Healy

Chapter 12: How Six Sigma Can Make Lean Even More Effective

The final topic that we are going to take a look at here is Six Sigma and how it can be used along with the Lean system. Many times you may have heard of Lean Six Sigma, or even Six Sigma on its own, and you may be confused about what the differences are. It is possible to implement the Six Sigma methodology on its own, and it is also possible to implement the Lean process on its own too. You can also implement them together in order to get even better results for your process and your business. Let's take a look at the Six Sigma methodology and how it can work together with the Lean process we have been discussing in this guidebook.

Six Sigma is known as the shorthand name of a system that measures the quality with an overall goal of getting as close to perfection in the process as possible. If a company is using Six Sigma properly, then they are going o generate as few as 3.4 defects per million attempts at the given process. Z-shift is going to be the name that is given to any deviations that are available between a process that was completed in a poor manner, and a process that was completed to perfection.

The standard Z-shift is one with a number of 4.5, but the ultimate value that businesses are aiming for is a 6. Processes that haven't been viewed with the lens of Six Sigma are going to earn about 1.5. A Six Sigma level of 1 means that the customers are able to get what they expect from the customer about 30% of the time. If the Six Sigma level is at 2, it means that about 70% of the time, the customers are able to get what they expect. If you can get to a Six Sigma level of 3, this means that about 93% of the time, the customers are satisfied with what they are getting from you.

If we move up the scale and get to a Six Sigma level of 4, this means that your customers are satisfied with the level of attention and the quality of the product they are getting more than 99 percent of the time. This means that reaching a Six Sigma level of 5 or 6 indicates that the satisfaction percentages with your customers will be even closer to 100 percent, or almost perfections.

In addition, you will find that the process of Six Sigma can be broken up into numerous certification levels. Each one is going to have a different amount of knowledge with it to help the individual know more about this process, how to implement it, and how to reach the near-perfection levels that are required with it.

The executive level of Six Sigma is going to consist mainly of management team members who are going to be in charge of going through your company and actively setting up the Six Sigma method. A Champion of Six Sigma is someone who is able to lead the projects you set up, and who will be the voice of these projects specifically.

There are also white belts in this system, and they are the rank and file workers. These are the individuals who have an understanding of Six Sigma, but it is going to be more limited than the higher two levels. The yellow belts are going to be active members on the Six Sigma project teams, and they have the responsibility of figuring out some areas where improvements can be used.

Next, are the green belts. These individuals work with the black belts on some of the projects considered high level, while also working to run some of their own yellow belt projects. Then there are the black belts who will run their own high-level projects while still doing some mentoring and some support for the other tiers we talked about. The Master black belts are going to be those who the company brought in specifically to implement this system in the business, and who can help to mentor and teach anyone, no matter what level they are at.

Implementing

Giving your team some compelling reasons to work with Six Sigma can be so important to how successful the whole process is. To make sure that the Six Sigma process is implemented in the right way, you need to find a way to motivate the team. Explaining the Six Sigma process to them and discussing how important it is to implement this new methodology can be a great place to start.

One tactic that works with this is to use a burning platform. This is a kind of motivational tactic where you are going to explain the situation that the company is in right now, and why that situation is so dire. Then you can explain that implementing Six Sigma is the only way to get long-term survival to last for the company. Of course, before you make these assertions, you should have some statistics that can help you to make this point.

Ensure that the Right Tools Are Available

Once your team has gone through the initial rounds of training that they need, it is important that you set up a kind of program for mentorship along with some additional refresher materials ready for any team member who may need them. You will find that at this stage, one of the worst things that can happen is for someone to be confused about one of the important parts of Six Sigma, and then they are rebuffed because they don't have the right resources to help them out.

The more information that you are able to provide to your team, and the more opportunities you provide for them to learn and understand Six Sigma in the beginning, the easier the implementation of this process will be. You will be able to see that everyone is on the same page, that everyone understands the importance of this process, and everyone has the right training and knowledge to make this process happen.

The Key Principles

Now that we know a little bit more about Six Sigma, it is time to take a look at some of the key principles. This process is going to work the best based on an acceptance of five laws. The first is going to be the law of the market. This means that before anything is implemented, the customer should be considered. The second is the law of flexibility. This is where the best processes are those that can be used for the greatest number of disparate functions.

Then there is the third law, which is going to be the law of focus. This one states that a company that follows Six Sigma should put all of their focus on the problem that they are experiencing, rather than on the business itself. Then there is the fourth law or the law of velocity. This one states that the more steps that are in a process, the more likely that some of those steps aren't needed, and that the process is less efficient. For the last law, we look to the law of complexity. This one states that the simpler the processes are, the more superior they are for the business.

So, how do I choose the best process? When it comes to deciding which of the processes that you should use with Six Sigma, the best place for you to start is with any process that you know to be defective, and that needs some work to reduce the number of deficiencies that occur. From there, it is simply going to be a matter of looking for situations where Takt time is out of whack before you figure out which steps where the number of available resources can be reduced as well.

The Methodologies to Work With

When you are ready to work with the Six Sigma process, there are going to be two main methodologies that you can choose to work with. Both are going to be efficient and can work, it just depends on which kind of industry you are in and what works the best for you. The two main methods are going to be DMAIC and DMADV.

First, we need to take a look at the DMAIC. This is an acronym that is going to help you and your team remember the five phases that come with it. This is useful when it comes to creating a new process and fixing any processes that need some extra work to be more efficient and deal with less risk. The way that DMAIC works includes:

1. Define what the process needs to do. To figure this out, you need to get some input from the customer and then work from there.

2. Measure: This is where your team needs to measure the parameters that the process will adhere to. Once this is done, you can ensure that the process is being created in a proper manner by gathering all of the information that is relevant.

3. Analyze: Here you will need to analyze all of the information that you have gathered. You may be able to see that there are some trends coming out of that information or find out that you need to do some more research and analyze it before continuing.

4. Improve: This is where the team can take that information and the analysis that you did, and make some improvements to the process.

5. Control the process. You need to work to control your business process as much as possible. You can do this by finding ways to reliably decrease how many times a delinquent variation starts to make an appearance in your process.

In addition to working with the DMAIC process, you can work with the DMADV process as well. This is very similar to the method that we just talked about before, and the five phases are going to correspond with the DMAIC process as well. The five phases that come with the DMADV method will include the following:

1. Define the solutions that you want the process to provide. You can look at your own mission statement, how the product is supposed to work, and input from the customer to help figure this one out.

2. Measure out the specifics of the process so that you are able to determine what parameters need to be in place.

3. Analyze the data that you and your team have been able to collect up to this particular point.

4. Design the new process with the help of the analysis that you have.

5. Verify any time that it is needed.

Both of these methods can be very effective at helping you to see results when you try to implement Six Sigma into your business. Often they work in very similar manners. You will need to consider the situations around your process, what deficiencies you need to fix, and more to help you determine which process is the best one for you.

Is Six Sigma the right choice for me?

While this process is something that can work for many different businesses across a wide variety of different industries, and Six Sigma has something to offer for teams of all sizes and shapes, it doesn't mean that this process is going to be the best fit for everyone. This can be really apparent as implementing it successfully means that a number of specifics need to come into play. This will start with the conviction of those who are looking to implement the system in the first place, as well as the overall culture that is found in the business and how open it is to the new change.

This is why many companies decide to ease into the process and will start with the 5S method that we talked about earlier. This is seen as a lower impact method that can adjust the team to what you want to happen before you move into some more advanced techniques, like what you find with Six Sigma. Once the team has accepted what you are trying to do, it becomes so much easier to implement all of Six Sigma and all of Lean into the business and its culture.

When you are taking a look at the Six Sigma method and trying to determine if this kind of transition is actually something that you can do, you must make sure that no one in the business, especially upper management, sees this as a fad or a trend that the business is just trying out. In fact, Six Sigma, and the whole Lean philosophy needs to be seen as an evolution of the ideals that the company already put in place.

In most cases, the more involved you can get the leadership of the team right from the beginning, the more onboard the team will be, and the more participation you will be able to get out of everyone. This is why it is so important to get all of the employees on board, whether they are in top management or hold another important position within the company.

In addition, it is so important for the culture of your company to be seen as one that is in full support of this kind of positive change, and to remember that if your upper management, or anyone on the management team, isn't able to come up with a consensus on the new program, that it is better to hold off a bit to reach that consensus. Jumping in when not everyone is on board, especially if some of those are the upper management, means that the idea and the process will be dead from the start.

Of course, this doesn't mean that every single person on the team must be committed to the idea of Six Sigma or the Lean methodology right from the start. But it does mean that the changes that occur need to be seen as institutional. This ensures that the front that you send to the public shows that everyone is united under the ideas of the method.

Implementing Six Sigma into your business can take some time, but when you add it together with the ideas of Lean, then you are going to see a big shift in the company culture and so much more. But when both of these ideologies are used together, you will find that it results in more satisfaction with your customers, less waste, more efficiencies in the process, and more profits in the long term of your business.

Conclusion

Now that you have come to the end of the book, I hope that you appreciate everything that Six Sigma stands for. Over the years, there have been a lot of myths and confusion regarding this particular methodology. However, one thing should be clear by now.

Six Sigma is extremely useful for any organization that wants to improve quality, reduce costs, and enhance the speed of delivery of goods and services.

You have learned the most important tools and processes that are used in Six Sigma implementation. Keep them in mind as you move on to the next phase of the journey – which should be implementing and deploying a Six Sigma project.

Remember to follow the right procedure when trying to identify a solution to defects in your business process. Use the DMAIC stages to guide you every step of the way.

Improve the organization and control over your line, process, area, department, shift or full factory starts with being clear about what you are trying to achieve.

You need good skills around you – enough strength in depth to deal with whatever is thrown at you. A team made up of skilled, experienced people who know what is expected and what to do is the first step in the journey.

Once the team has enough key skills, it will have the capacity to deal with the day to day issues and challenges that occur in factories whilst being able to take on adopting new practices such as 5S.

5S will clear the decks of clutter, make the essentials like tools easier and quicker to locate and will give the area an organized and controlled appearance and "feel".

Keeping on top of regular red-tagging, sorts and sweeps will allow you to maintain this level of organization. You can't do it only once! It must be done until it is so habit-forming that the team do it without even thinking about it. This can take some time but it will happen if you stick with it.

Visual lean techniques like shadow and line marking will reinforce the look and feel of a controlled organized workplace.

SIC and SPC are further tools to bring in place to allow the team to quickly see and understand where they are against where they need to be.

Throughout keep using a DMAIC approach to keep on track, keep communicating to your team and everyone involved. This will create a positive "feedback" loop by allowing people to see the improvements flowing through. Seeing improvements being delivered creates a feeling of positivity and spurs teams on to achieve more.

Keep moving forward, don't quit even when things don't happen as well or as fast as you want them to, and before you know it you'll be seen as a someone who can deliver improvements and change within your business.

CPSIA information can be obtained
at www.ICGtesting.com
Printed in the USA
BVHW040302211020
591325BV00014B/738